The Holistic Egyptian Quality Management Approach “HEQMA”

The Holistic Egyptian Quality Management Approach "HEQMA"

By

Osama Kamal, PhD

CAMBRIDGE SCHOLARS PUBLISHING

The Holistic Egyptian Quality Management Approach "HEQMA",
by Osama Kamal, PhD

This book first published 2013

Cambridge Scholars Publishing

12 Back Chapman Street, Newcastle upon Tyne, NE6 2XX, UK

British Library Cataloguing in Publication Data
A catalogue record for this book is available from the British Library

ISBN (10): 1-4438-4668-6, ISBN (13): 978-1-4438-4668-4

TABLE OF CONTENTS

LIST OF TABLES

ACKNOWLEDGEMENTS

I am grateful to Professor Lotfy Louise, my supervisor, who handles a student with a free spirit like me and my project with the right touch. His guidance and encouragement were so important to me.

For Professor Hasan El-hofy and Professor A. El-Kashlan, their ideas had an impact on my approach, for Professor Sherif El-Araby, to whom I have turned repeatedly regarding facilitation of the project. I pay tribute to them all.

I would like to extend my appreciation for my examiner; Professor M. Anwar El-Dardiry and his patience in reviewing this thesis and for his comments.

I would like to extend my appreciation also to the following organizations: production engineering department, Alexandria University and the Arab academy for science and technology and maritime transport.

Industrial experts from the following companies: El-Araby home appliances, El-Araby for lighting Technology, Mr. Samuel Beshay, EFQM representative in Egypt – Alexandria Petroleum Maintenance Co. Petromaint; Ex-chairman and board members. Amria pharmaceutical Co.; Pharco, Pirelli Tires Co; Egyptian Petrochemical Co.

They assisted me in conducting the focus group in different industries.

To my sister, the doctor whose enthusiasm for certain ideals has made a great impact on me, to Dr. Soliman Gliwan, my Libyan colleague in the PhD journey; they offered their endorsement.

Talking about research is one thing, actually doing it is another. As a student who was fond of quality management work in the nineties of the past century, the road is far from straight. I realized the impact of the dual people-process concepts on the perceived quality model, and that this simple structure was too pretty to be dismissed. In the end, all should be left to practical applications.

Osama Kamal

E-mail: OKamalabd@hotmail.com

LIST OF ABBREVIATIONS

BSC	Balanced Score Cards
BPR	Business Process Reengineering
CSI	Satisfied Customer Index
EIO	Egyptian Industrial Organization
EMS	Environmental Management System
EQMA	Egyptian Quality Management Approach
EFQM	European Foundation of Quality Management
ISO	International Organization for Standardization
JIT	Just in time
MBNQA	Malcolm Baldridge National Quality Award
MPI	Measuring Performance Improvements
NQI	National Quality Institute
OHSAS	Occupational Health and Safety Assessment Series
OJT	On Job Training
P&QI	Productivity and Quality Inst
QMS	Quality Management Systems
RADAR	Review-Assess-Deploy-Approach-Results

CHAPTER ONE

INTRODUCTION

Industrial organizations in developed countries have passed many quality progress phases. There was quality control, Q.C., passing to quality assurance, QA, and reaching TQM. Egyptian Industrial Organizations, EIO, bypassed or jumped over the Q.C. phase in many cases, to practice Quality Management Systems, QMS, without verifying product quality first.

From a historical stand point, the EIO was compelled to apply different QMS in the last decades, although they belong to the sixties of the last century.

On the other hand, there are increasing concerns of improving organizations' performance levels in the developing industries, and the recognition of the influence of human factors on performance.

So, during the last decades, many quality management entries have been coined to include in an excellent organization's characteristics: the Human Resource capital.

The quality is as important as either a system or characters. Problems in the Egyptian Industrial Organizations (EIO), such as quality control problems, have led to moderate product quality and management problems. So there is a need to concentrate on quality management models that take care of Q.C. principles in order not to miss an important component of global quality constituents.

As good remedy efforts for a Quality Management project, it is necessary for the Egyptian organization's top management to be fully aware of the concepts and requirements of excellence management. Top management should realize the importance of international applications and initiatives of specialized agencies in Measuring Performance Improvements (MPI) models for superior quality.

Those management initiatives are also useful in the self-assessment process. This research focuses on those concepts later.

The EIO needs, in this respect, to experiment with a Quality Management logic which is close to its industrial business culture in the current phase, instead of applying scattered and diversified Quality models which can waste time and effort.

On the other hand, it cannot be ignored that many organizations have rightly been guided to searching solutions for a new organizational formula to rise with and reach the status of a "high performance" organization.

1-1 Research Problem

1 - Why is the implementation of QM models not achieving targeted results?
2 - Can we find a model that is derived from TQM, which is ever fertile and promising; to explore new integrative management ideas that suit Egyptian industrial culture?
3 - What are the methods for measuring performance improvement in EIO?
4 - With the validity of the new TQM approach as a national one, can the profit of EIO market and the field of QM be improved?
The absence of the domestic high technology for upgrading the quality of our products, along with cultural dimensions, can call for a cultural scale.

According to the Hofstede scale, the Egyptian characteristics in particular of the Arab world have the predominant characteristics of "Large power distance" and "Uncertainty avoidance", he also pointed out that the loyalty in a "collectivist society culture" is paramount; (such as obedience, group spirit, social relations, etc.).

Other characteristics he uses are long term orientation and masculinity (geert-hofstede.com.3/2009).

The above scale will be considered in the formulation of the proposed model. For example, Leadership is an opponent to the concept "Large power distance".

In historical order, Egypt had many charismatic leaders many decades ago, such as Talaat Harb Basha, Aboud Basha, and Farghaly Basha who fought for national belongingness and vision. The appearance of such role models is encouraging for grouping team work and agreement on conjoined goals.

The importance of role models is to enhance the leadership for the application of a good QM model.

The problem is that no one has previously engaged boldly to intervene and tackle the situation of developing new approaches for "HOW" to improve performance.

1–1–1 Pilot study

The pilot study was done to document that there is a problem with the existing TQM models being applied successfully in the Egyptian business arena. Content validity of the research is achieved through this pilot study.

There is a relationship between optimal Quality Management Systems and the creation of domestic high technology; the study also looks at this aspect.

The study was done through a review with three academic members from the Arab Academy for Science and Technology and Maritime Transport AASTMT and Alexandria University to adjust the focus group theme and problem. At the same time, there were initial interviews with a group of top management of Petromaint, Projects GM, Proposals and contracts GM, and the financial GM, who is a board member as well. Interviews were also held with the chairman (PHD holder) and his assistant (MBA holder) as "key informants". This has been very helpful in developing and designing HEQMA focus group dimensions and questions, as they have professional expertise consolidated by academic background.

A focus group is a new analytical method which uses a lower number of samples than a mail questionnaire, but with much greater depth in the information collected and in reliability, as it depends on face to face interviews. As Aoki (2007) declared, the divergence between canonical and non-canonical practices suggests that actual practice cannot be controlled completely through formal procedures.

The field study helped to enhance the knowledge and ideas of the researcher toward the research subject and its results. It also guided the rephrasing of the subtitles, management ideas, and concepts used in EIO. Interviews helped with the procedures of the consequent focus group and on getting the proper communication channels with relevant authorities.

The problems with EIO call for reaching a quality management model suiting its circumstances, sourced from ground experience, and verified by Egyptian industrial expertise.

It is concluded then that there is a gap between what Egyptian manufacturers' demands are, and what they get regarding international quality management models.

Another broad problem area was expanded through literature reviews, which revealed the need to develop a new national model. The researcher also expanded the broader problem area through practice.

1-2 Research Hypothesis

Among the existing quality management concepts that have relations between salient constructs, the main TQM variables presented in the proposed model are gathered in 4 bundles containing the ten main items of the model.

- Independent Variables are: Top Management Commitment and Corporate Ethics
- Dependent Variables are: Outcome factors: Marketing Results and Satisfied Customer Index CSI
- Moderate variable is: Human capital: Motivation, Team performance, training and development
- Intervening variable: Internal capabilities (Improving productivity, Cost of Quality, Safety and positive Impact on Society)

Research hypothesis are:

H1: Egyptian Top management commitment has a direct and positive impact on their Customer satisfaction.

H2: Egyptian Corporate Ethics have a direct and positive impact on the HR capital.

H3: Top management commitment has a direct and positive impact on their organization's internal capabilities.

H4: HR Capital has a direct and positive impact on the organization's marketing results.

1–3 The links between variables

To illustrate the philosophy of building up this Egyptian model, it is fundamental to say that the key role of top management is to create corporate values and vision, and inspire people by active positive attitudes and behaviours; those are rooted in ethics.

This will lead to envisioning and establishing cooperative teams and a learning organization spirit through training and people development. Thus managers can manifest the internal capabilities of the organization, which will stimulate and interact with quality marketing, and project them in a clear and consistent manner to raise the satisfied customer index (CSI).

It can be seen then that the model is dynamic and has consecutive relations between inputs and outputs, with intermediate and intervening factors working as catalysts.

1-4 Research Objectives

1 - Finding the best modules that fit for Egyptian Industry, thus improving the competitiveness of the EIO in particular and eventually the Arabic practice of TQM in general.

As Evans (2005) expressed:

> "Measuring organizational success and implementing effective strategies for the future represent continuous challenges for managers, researchers and consultants".

2 - Formulating a Holistic Egyptian Quality Management Approach **HEQMA**.

3 - This will eventually strengthen the progressive industrial quality, moving it one step up, eventually supporting the integration of the Egyptian economy into developed industrial alliances.

4 - Another aspect of importance for this study is the development of an applicable quality management model which can be used by the research community, or practical applications in other dimensions. In the academic paradigm, this research would unfold the subject of developing new quality management models for specific sectors in the EIO. The topic is expected to unleash the horizons for introducing practical but scientific solutions that the Egyptian businesses are thirsty for.

5 - In essence, the results of this research can help EIO management in making crucial decisions and in resource allocations that are required to make the HEQMA implementation a success for a group of them.

1-5 Research methodology

It is important to highlight that the research area should develop many questions on management culture for the benefit of the Egyptian industry; this research used many methods to achieve its objectives:

1 - Reviewing the literature on the current TQM status and success factors for implementation, worldwide and in Egypt.

2 - Capturing the advantages of each major international quality management model over the others, and using their harvested benefits.

3 - Evaluating how HEQMA can be successful in terms of EIO cultural dimensions. Moreover, the study gets closer to the Egyptian industrial experts, academics, and representatives of international QM models to grasp and use their knowledge on HOW to improve EIO.

4 - Using descriptive analysis to capture the relationship between TQM concepts and successful implementation.

5 - Developing a model and scoring system for a TQM approach suitable forlidation by a Focus Group technique amongst Egyptian industrial managers, representatives of International QM models, and academic scholars; qualitative data is collected to assure a salient "HEQMA", which accords the Egyptian business domain.

For the first model of variance, for the 10 proposed model items, Cronbach Alpha was found to be -0.756.

For the second model of variance is for questions and interview, Cronbach Alpha = 0.7514.

This shows that the two models of variance are stable and the interview questions are reliable.

It can be seen that the conclusion of the research is in accumulative portions; not necessarily to be a sudden shock or fundamental change in concepts.

1-5-1 Literature review

Different maturity of Quality forms have been reviewed; philosophies, models, and tools. Their current application in Egypt was then reviewed.

Many Quality models are discussed in the following chapters. However, to sustain reliability of the research, a clear definition and methods of each quality model is demonstrated, including reasons for application, success, and potential obstacles.

Quality models are then judged and compared, including models developed by other researchers. Links and gaps between those models are pointed out.

During the construction of this model, an explicit relationship between the various TQM tools and concepts is established. There is value in looking at customizing some requirements of the international quality standards as a dependable source for quality sub-items. Also, quality dimensions that are not necessarily a requirement of a standard, but work as a cultural catalyst, are integrated in the proposed model, such as cost of quality and some entries of training and corporate ethics. This approach can then be a model for how to measure the performance improvement (MPI).

A vision for the future is gathered then, out of the literature reviews to form the imposed model.

1–6 Research Outcomes

- Identifying HOW TO MEASURE and the "mechanisms" to improve different quality elements in detailed attributes.

The model takes care of the quality control elements in Egypt, where they have been ignored for decades; the model is not ignoring the Human Resource capital of the Egyptian people. The model linked goals, measures, and improvement activities all together. For this, the model elements have to run simultaneously.

- Defining the main quality elements needed for Egypt in four bundles, while integrating them in a unique way.

- Calling for other business attributes to be included into quality concepts, such as financial, safety, and marketing perceptions.

HEQMA specifies where the organization should consider correcting its path, and then improve its results, as well as the result indicators for excellence to be achieved.

- The goal of this research is to strengthen the Egyptian industrial quality, moving it one step up, and eventually supporting the integration of the Egyptian economy into developed industrial alliances.

1-6-1 How the model will benefit Egypt and the Arab world

The HEQMA model helps top management to reach business excellence without the need for a management consultant, or at least it saves half of the consultancy service.

Theory and application: Theory and application were linked in HEQMA by numbers of ways: through the stages of the research project, many industry leaders and consultants were investigated. Also, linking demonstrated quality entries with HEQMA while discussing them.

Also, looking for quality management models as a certification or award only will not be a good business model, or an academic research project.

In HEQMA, there has to be human leverage, training on different soft skills, and systematic experiences devising, through succession, plans for example.

CHAPTER TWO

LITERATURE REVIEW

Using different quality management approaches benefits organizations differently. In this chapter, each quality management model or technique will be discussed to create constructive points for their understanding and use. The demonstrated concepts here represent a base for the QM model formulated later in this research study.

2-1 International QM Philosophies, models and tools:

To be specific, it is intended to review the different quality maturity levels:

1 - There are integrative quality gurus such as Deming, Crosby, Juran, Feigenbaum, and others; there are also the generic approaches which can be classified as a Philosophy, such as TQM, Business Process Reengineering BPR, and World Class Manufacturing WCM.
2 - On the other hand, there are models like Balanced Score Cards BSC, Malcolm Baldridge National Quality Award (MBNQA), European Foundation of Quality Management (EFQM), ISO9000, and others, which can be considered as the quality management system models.
3 - The remaining tools and techniques include Kaizen, Just in time JIT, Lean Production, Six Sigma, and Bench marking.
The following is a demonstration of the above. To sustain reliability in discussing the above, a clear identification is demonstrated during its presentation, and distinctive edges and any obstacles will be highlighted. Also, the operating conditions and actuation circumstances will be investigated where possible.

2-1-1 Edwards Deming

As a major quality guru in recent decades, the following are a condensed list of the 14 points regarding Deming's management:

1. Create constancy of purpose toward improvement of product and service.
2. Adopt the new philosophy.
3. Cease dependence on inspection to achieve quality. Eliminate the need for inspection on a mass basis by building quality into the product in the first place.
4. End the practice of awarding business on the basis of price tag. Instead, minimize total cost. Move toward a single supplier for any one item, on a long-term relationship of loyalty and trust.
5. Improve the system of production and service constantly and forever.
6. Institute training on the job.
7. Institute leadership; the aim of supervision should be to help, and use machines and gadgets to do a better job.
8. Drive out fear so that everyone may work effectively for the company.
9. Break down barriers between departments. People in research, design sales, and production, must work as a team.
10. Eliminate slogans, exhortations, and targets for the work force asking for zero defects and new levels for productivity.
11. Eliminate work standards (quotas) on the factory floor. Substitute Leadership. Eliminate management by objective. Eliminate management by numbers or numerical goals.
12. Remove barriers that rob the hourly worker of his right to pride of workmanship.
13. Institute a vigorous program of education and self-improvement.
14. Put everybody in the company to work to accomplish the transformation.

The above principles are a mix of Q.C., assurance, and TQM concepts.

2-1-2 Feigenbaum principles

1 - Total Quality System.
2 - Design Quality in.
3 - Customer Orientation.
4 - Total quality Control.
5 - Quality is a company-wide process.
6 - Quality is what the customer says it is.
7 - Quality and cost are a sum, not a difference.
8 - Quality requires both individual and team zealotry.

9 - Quality is a way of managing.
10 - Quality and innovation are mutually dependent.
11 - Quality is an ethic.
12 - Quality requires continuous improvement.
13 - Quality is the most cost-effective, least capital-intensive route to productivity.
14 - Quality is implemented with a total system connected with customers and suppliers.

These quality awareness principles are a good prerequisite for any quality movement.

2-1-3 Crosby principles

Crosby (1979) mentioned that a quality vaccine is: Determination, Education and Implementation. His fourteen steps to quality improvement are:

1 - Evolve from the need to define, understand and communicate the "absolutes" throughout all levels of an organization. This is a methodology for implementing the quality improvement process (disliking the word "program", as it implies temporary) in an organization.
2 - Ensure that clearly visible signals that management is committed to quality permeate throughout the organization.
3 - Setup quality improvement teams with a senior representative to demonstrate high-level commitment.
4 - Measure and identify current and potential quality problem areas.
5 - Evaluate "cost of quality" and ensure its potential as a management tool is understood at the appropriate levels.
6 - Raise the level of quality awareness and encourage it as a personal concern of everyone.
7 - Ensure corrective actions are placed, the importance is accepted and they are adequately addressed.
8 - Establish a monitoring system for the improvement process.
9 - Educate and train supervisors on their role and responsibilities in the quality improvement process.
10 - Program "Zero defect days" to communicate progress to all levels, taking the opportunity to reaffirm management's commitment to quality.
11 - Encourage that which allows individual and team goals to be set.

12 - Encourage two-way quality communication and highlight obstacles to achieving individual and team goals.
13 - Recognize and appreciate individual and team efforts.
14 - Establish quality councils to discuss quality matters on a regular basis.

Visibly repeat steps 1 to 13 to demonstrate that the improvement process is continuous.

This is another wave of inculcating quality awareness in the organization.

2-1-4 Juran principles

As mentioned in Juran (1993), are:

1 - Management involvement.
2 - Quality planning.
3 - Quality control.
4 - Quality improvement; let Quality becomes a part of each upper management agenda.
5 - Quality goals enter the business plan - Goals are deployed to the action levels.
6 - Measurement is established throughout the organization. Upper managers regularly review progress against goals.
7 - Recognition is given for superior performance.
8 - The reward system is revised.
9 - Benchmarking.
10 - Training is done at all levels.

Top management's commitment is clearer in Juran principles; MPI is now a mature concept. Juran also claims that Philip Crosby's Zero Defects approach does not help, since it is mistakenly based on the idea that the bulk of quality problems arise because workers are careless and not properly motivated.

Benchmarking cannot be utilized properly by EIO.

Further to this, it may be beneficial to mention the philosophy of the Productivity and Quality Institute P&QI and Ishikawa in the same context.

2-1-5 Ishikawa principles

Ishikawa sees the Company-wide Quality control movement as implying that quality does not only mean the quality of product, but also of after sales service, quality of management, the company itself, and the human being. This has the effect that:

1 - Product quality is improved and becomes uniform.
2 - Defects are reduced.
3 - Reliability of goods is improved.
4 - Cost is reduced.
5 - Quality of production is increased, and it becomes possible to make national production schedules.
6 - Wasteful work and rework are reduced.
7 - Technique is established and improved.
8 - Expenses for inspection and testing are reduced.
9 - Contracts between vendors are rationalized.
10 - The sales market is enlarged.
11 - Better relationships are established between departments.
12 - False data and reports are reduced.
13 - Discussions are carried out more freely and democratically.
14 - Meetings are operated more smoothly.
15 - Repairs and installation of equipment and facilities are done more rationally.
16 - Human relations are improved.

The Ishikawa principles are also focused towards Q.C. in a clear manner.

2-1-6 The philosophy of the Productivity and Quality Institute

As one of the Arab Academy for Science and Technology's units, the P&QI has a strategy of:

1 – Understanding, documenting and reviewing customer requirements and responding accordingly;
2 – Adopting a philosophy of continuous improvement;
3 – Attaining the most recent international standards in different fields;
4 – Updating scientific and study courses to the most developed worldwide;
5 – Utilizing the most competent and qualified human resources and continuing to develop their skills;

6 – Conducting peer service agreements with big expertise houses worldwide.

The above are the main Gurus' principles which are integrative and cover almost all quality concepts, with more details. The first four quality gurus were consulted regarding the integration of their principles into the proposed model, where applicable:

2-1-7 Juran principles

The first principles were stressed in this model: management involvement is achieved through many indicators of the model regarding management commitment. Also many items reflect quality improvements in the form of percentage improvements, as is clear in the model. Long term planning and benchmarking may be difficult to achieve in Egypt due to the varying industrial circumstances.

2-1-8 Deming principles

The essence of his principles where adopted.

2-1-9 Crosby principles

They are taken into consideration when establishing the model sub-items. All principles are included in the model items except for the zero defects principles, which may be far away, but can be a future strategy eventually.

2-1-10 Feigenbaum

The model is a system for total quality; it reflects not only quality for customers, but profit for all stakeholders and good income for employees. The customer orientation is clear in the system in many ways of the satisfied customer index.

The Leadership, which is a fundamental component of TQM, is transformed in the model in the form of "Top management commitment" and "corporate Ethics" to suit the collectivistic culture of Egypt.

2–2 Quality Approaches

From the above, many quality approaches and business processes are generated; amongst them are:

2-2-1 Business Process Reengineering (BPR)

BPR philosophy is a major or root change of the organization and is based on radical rethinking. It should not be looked at as a mending or repair process, but as a new arrangement or disposition to overcome redundancy. Since its spread through the literature in 1993, the BPR concept became one of the modern theories through which the specialists found a way to confront organizational collapse and market share loss. BPR advocates the customer-driven mechanism.

Stebbins et al (1998), explains the mechanism established within the organization for organizing and conducting the BPR change program, and that it determines to a large extent the capacity to carry it out and its success. The nature of the mechanism's structural configuration, its creation, key processes, and linkages to the formal organization, also determine employee commitment to change.

Management experts on BPR have proposed that BPR consider the integration of all aspects of the system while focusing on the reengineering of key processes. Paradoxically, the focused energy and effort on reengineering key business processes results, in many cases, in a lack of attention to integration. At issue is whether BPR change initiatives are narrowly focused versus comprehensive system-wide change programs.

Tennant and Wu (2005) pointed out that the concept of business process improvement encouraged businesses to consider company-wide processes rather than focus on production processes only. This led to the conception of business process reengineering (BPR) to promote "the fundamental rethinking and radical redesign of business processes to bring about dramatic improvements in performance". BPR remains the quest for improvement through quick and substantial gains in organizational performance by starting from scratch in designing the core business process. This has led to BPR being promoted as one of the major modern techniques of changing management within organizations.

However, O'Neill and Sohal's (1998) study showed that *the Organizational benefits from BPR*, at least to a moderate extent, were:

- Productivity, decreased cycle time, inventory or cost;
- Profitability, increased economic growth; and

• Quality, improved products or services and related information.

The same study showed that the most severe problems in BPR implementation were in the minor to moderate level BPR problems. Problems were then classified broadly into the groupings of planning, operational, up-front costs, side effects, organization environment, and problems due to a lack of results from BPR projects.

Al-Mashari et al (2001) declared in their study findings that the six change management elements in BPR were: Roles and responsibilities, organizational structure, management systems, skill requirements, shared values and beliefs, and reward and recognition systems.

The proposed model in this research supports the measurable and gradual corrections and improvements rather than radical changes the BPR brings.

2-2-2 World class manufacturing (WCM)

The phrase "world-class manufacturing" is one of the most overworked terms in management glossaries at present. However, one of the most comprehensive definitions of WCM is provided by Salheldin and Eid (2007), which refers to many techniques and technologies designed to enable a company to match its best competitors. These techniques include for example, JIT, quality circles (QC), Kanban, material requirements planning (MRP), flexible manufacturing system (FMS), computer aided design (CAD), computer aided manufacturing (CAM), computer integrated manufacturing (CIM), manufacturing resource planning (MRPII), total quality control (TQC), total productive maintenance (TPM)/Preventive Maintenance, TQM, simultaneous engineering, benchmarking, intelligent manufacturing, electronic commerce, business process re-engineering (BPR), enterprise resource planning (ERP), electronic data interchange (EDI), and supply chain management.

The same reference added that WCM has emerged as a result of many business drivers. Firstly, the changes in the driving forces for manufacturing strategy, from an initial push to improve current business processes to achieve savings and improve efficiency, the companies have come to be driven by a desire for greater supplier involvement and customer service in later implementation which lead to the adoption of a mass customization production philosophy. Secondly, competitors' use of the WCM techniques and response to customers also has a strong effect on the adoption of the most advanced WCM techniques for the production purposes. Thirdly, as a result of its growing ability to bring new opportunities and to facilitate the development of the new organizational forms and structures needed to meet the continuously emerging changes in business imperatives, the WCM importance increases as it becomes involved in each task in today's business. Finally, IT developments are also forcing organizations to be up-to-date in their use of advanced technologies regarding delivery of speedy and high quality information, as well as facilitating greater degrees of communication and integration across business units and external partners.

Drivers and barriers to WCM implementation in general have been classified in various ways; for example, regarding external drivers: global competition, international customers' needs, developments in IT. Internal drivers include changes in the organizational strategies and savings. WCM barriers can be the need for cost justification, resistance to change, lack of management support, lack of knowledge, lack of appropriate monitoring, and lack of employee education and training.

It is clear from the researcher's point of view that WCM is a class that is a long way from being adopted in Egypt for the time being.

2-2-3 Total Quality Management (TQM)

When conducting a study on TQM, it is recommended to highlight the ambiguity associated with the term, it is also important to clarify what TQM refers to. If we go back to the origin of TQM, the definition given by the British standard BS7850 is as follows:

TQM is a Management philosophy and company practices that aim to harness the human and material resources of an organization in the most effective way to achieve the objectives of the organization (BS7850, Part1, 1992).

Some fundamental concepts for TQM, as per the same reference, are:

1 - Commitment
2 – Customer satisfaction (Internal & External)
3 – Quality losses
4 – Participation by all
5 – Process measurements
6 – Continuous improvements
7 – Problem identification
8 – Alignment of corporate objectives and individual attitudes
9 – Personal accountability
10 – Personal development.

TQM considers that applying a quality system is an essential supporting technique for TQM.

Nevertheless, El-Araby and Abdel Ghany (2007) have carried out a literature survey which provides guidance on the most authoritative reading on the subject. They noted that:

> TQM has been defined as an approach, framework, methodology, philosophy, process, system…etc. by various researchers. While a number of empirical studies reached the conclusion that its implementation improves organizations' operational and financial performance, other studies claimed a negative relationship among same, whereas still others denied any effects.

In the researcher's view, those who denied the effects of TQM may need to apply BPR instead.

The same reference also pointed out the most significant factors that have contributed to the persistence and strength of TQM, which are:

The recognition and demonstration of the importance of quality as a source of superior competitiveness;

The success of Japanese firms in taking and retaining a market share from their Western counterparts;

The Influence of the teaching and writings of the so called quality gurus, and the introduction of recognized quality awards.

In the personal interview with Dr. El-Araby, in the pilot study phase, he explained that Deming has lately reached the theory of knowledge; that the CEO of the organization provided a crew of specialized experts in every branch he dominates. Dr El-Araby added that TQM is inculcated in all the other quality entries: JIT, 6 sigma, kaizen, lean, etc., and that the TQM relation with HR is diversified, for example empowerment has about 70 dimensions; concerning involvement and participation, a subsidiary of it is to study types of teams at work.

It is not sufficient to inspect, control, or assure quality in order to achieve customer satisfaction. TQM requires the application of quality management rules and principles to every component and at every level of the organization. Everyone should be committed to continuous improvement in their part of the operation. Through this participation and commitment, with the use of different tools and techniques that the TQM concept has adopted or developed, quality can be managed effectively (Eraqi 2006).

However, Raisinghani et al (2005) found that for TQM implementation, in many cases, it is followed by disappointment because implementation was not comprehensive enough to allow all the changes to take place.

These changes are both political and personal within the organization. In some cases, leadership development for management is necessary for them to act as change agents advocating TQM.
The researcher endorses the author regarding this.

2-3 Quality management Models

Quality models are the second level of quality understanding and inputs, which are basically derived from the above mentioned approaches and philosophies.

2-3-1 Toyota Production System TPS

The study will promote emphasizing both Kaizen & TPS.

As Aoki argued, in Japan there is a trend that companies in various industries try to implement kaizen activities in line with the Toyota way.

Therefore, it is possible to consider kaizen activities in Toyota as representative of Japanese kaizen activities.

From a different view, some literature sees The Toyota system called lean production (LP) (Olivella et al, 2007).

Nevertheless, it is worth saying that it is difficult for the TPS to be imitated by others, because it consists of a lot of interdependent systems (Teece et al., 1997, cited in Aoki, 2007).

The TPS has at least two main pillars: just-in time (JIT) and Jidoka or "automation with a human touch" (Elsey 2000). The former refers to inventory management; that is, practical ways and means of ensuring that at each manufacturing and assembly stage in the production chain only the correct number of products are supplied when needed on time. While the latter refers to responding to the needs of the human body and brain, i.e. ergonomics.

Also, Elsey pointed out that Kaizen, as it is about quality assurance, supports the TPS in two main ways. The first is through the formal emphasis on "groupism" (a Japanese term) and teamwork (similar to the sub-item "team work" in the proposed model of this research); the second way in which kaizen exerts an influence is to expect employees to work with conscious awareness of the importance of quality assurance and solve problems quickly (Elsey2000).

2-3-2 Balanced Score Cards BSC

Now we land to a rich quality practice which the researcher found more suitable to advocate the customized implementation of in Egypt. The BSC was developed by Kaplan and Norton in 1992 using large organizations as the basis for the model. The researcher suggests that the four dimensions of BSC are more applicable to EIO as well.

Essentially the BSC model consists of four quadrants, each containing different types of organizational goals, measures and improvement activities (Financial-Customer-Internal processes-Learning & Growth). Kaplan and Norton showed in 1996 that all four quadrants are dynamically interlinked when the BSC is used to develop TQM-based change programmes (Kapalan and Norton, 2004).

As Gautreau and Kliener (2001) explained:

> BSC approach generally groups strategic measures into one of the following: outcome or driver measures, financial or non-financial measures, and internal or external measures.
>
> Outcome measurements indicate the result of a strategy e.g. improved quality, higher revenues. Outcome measures, like profits, are indicators of how successfully the company has implemented strategy and are thus 'lagging indicators'. Conversely, driver measures are 'leading indicators' (like cycle time).
>
> Driver measures indicate incremental changes that ultimately affect outcome. BSC focuses on driver measures, since they are indicators of things to come, however outcome measures are still a vital part of the approach.
>
> Financial measures have been the most widely used performance measure in the past. Examples would be return on equity (ROE), profit margin, etc. Companies have developed sophisticated systems to help measure financial performance. These systems, however, do not measure non-financial performance, which is an area of increasing importance'.

Companies must look at both types of measures when determining strategy. An internal measure may indicate that a company is performing well, while the external measures show poor performance. Shrinking defect rates may be in accordance with internal strategy, yet the company could be viewed negatively by the market, resulting in a falling stock price. This may be a sign indicating that the internal and external strategies should be re-thought and/or re-aligned.

Steps to Make BSC Work: According to the same reference as above, the balanced scorecard requires that a company strategy be defined. The scorecard does not define the best strategy for a company to take. No system can do that; it is senior management's responsibility and vision. The scorecard also cannot select the best measurements of strategy. That is also senior management's responsibility, as well as lower/middle management's responsibility. The scorecard does require management to focus on creating strategy and defining ways that performance can be measured in accordance with strategy.

The original authors of the balanced scorecard recommend that the scorecard be up-dated on a regular basis; it is a dynamic model. The scorecard needs to be revised when changes in corporate strategy and/or corporate structure occur.

Managers must be clear and concise in communicating the strategy and in setting the performance goals for employees.

The scorecard must be constantly up-dated. This is good because it requires re-alignment with changing strategies or corporate structure. This also has a negative impact. It takes a great amount of time and resources to keep the scorecard updated and effective.

Measures may be difficult to quantify and the approach can lead to too many performance measures.

The Balanced Scorecard at Work: A generic framework for the balanced scorecard approach was presented by Kaplan and Norton; however, each company should have its own version of the scorecard. Different companies should have different measurements, which are important in achieving the company strategy. The most commonly used measurement categories are financial goals, customer satisfaction, employee satisfaction, productivity, and growth and innovation.

Organizations using the BSC often cascade the initial corporate level BSC down to several levels of linked scorecards, even to personal level for personal appraisal (McAdam1999).

Concentrating on the non-financial measures will also benefit the organization in the long term. As Evans (2005) described it, the financial measure is the history of a firm. This literature debated whether performance indicators drive the strategy or vice versa.

The financial and customer quadrants are integrated in the HEQMA outcome factor.

The internal processes and learning and growth quadrants are represented in the HEQMA internal capability and human capital factors respectively.

The difference in HEQMA is the "cultural" dimension; the base factor: corporate values demonstrated basically by the top management.

That basis already exists in the American industrial culture; it is added here in HEQMA as a new attribute.

There are a few items that managers should keep in mind when implementing any type of performance measurement system:

1) *Focus on a few key measures.* Too many measures may blur the strategy/goal.

2) *Use measures that employees can control.* If they cannot control the measures, employees will feel alienated and disgruntled.

3) *Keep the lines of communications open and revise the programme often with changes in corporate structure and strategy.*

Using some helpful hints like these will help in implementing a balanced scorecard or any type of performance measurement system (Gautreau and Kliener, 2001).

2-3-3 European Foundation for Quality Management (EFQM)

The success of the Baldrige Model (USA) and the Deming prize (Japan) encouraged the formation of the European Foundation for Quality Management (EFQM) in 1988. The model's criteria, sub-criteria and potential areas to address were given through a major revision in 1999 (Park, 2008).

Changes that took place in 2010 astonishingly matched many HEQMA items which were predetermined in 2009 (www.efqm.org, 5-2010).

The direct and clear links between EFQM 2010 and HEQMA proposed in 2009 are: HEQMA philosophies, Stakeholders' interest (1c above), understanding PEST effects (2a above), and financial consistency of the model (4b above).

For HEQMA base elements, ethics is highlighted (as in 1a above).

For an HEQMA indicator, there is the decision making process (1e above).

The Self-Assessment process allows the organization to clearly discern its strengths and areas in which improvements can be made, and culminates in planned improvement actions which are then monitored for progress.

Thus, it can be seen from the above definition that self-assessment is a vehicle for systematic continuous improvement in an organization (Hides et al, 2004).

After EFQM was introduced, it is worth saying that the system is based on the RADAR cycle, which is similar to the Deming cycle. The same reference pointed out that, to help organizations with the process of self-assessment, RADAR logic was introduced. RADAR is an acronym for Results, Approach, Deployment, Assessment, and Review (Hides et al, 2004).

As was demonstrated above, the EFQM is based on Self-Assessment. It is suitable here to emphasize the meaning. Williams and Dale (2000) have done a reasonable study on this.

It is important to train the managers and employees of the units where the first self-assessment pilot(s) will start. At this stage not every management team will be enthusiastic about the self-assessment activity, so it is useful to start with the ones who are the most supportive, or the ones that can be stimulated most easily. It is also important to invest sufficient resources, to give managers and employees the necessary time required for the self-assessment activities. In the first attempt at self-assessment, it takes some effort to gather the required data. The method and approach of assessment will also have to be selected and agreed.

To present EFQM weights, which are extracted from www.efqm.org (5/2010), the enablers & results modules of the model are:

ENABLERS: Leadership (100 pt), People management (90 pt), Strategy (80 pt), Partnership and resources (90 pt), Processes (140 pt);

RESULTS: People results (90 pt), Customer results (200 pt),

Society results (60 pt), Key results (150 pt).

2-3-4 ISO9000 Quality Management Systems

ISO9000 are considered the most common style of quality management system as guidelines for enterprises in Egypt.

According to the ISO organization in Geneva, more than one million companies have gained certification worldwide (www.iso.org 3/2010). As Boys et al (2003) claim, there are numerous reports that describe quality management practices and the benefits that emanate from implementation of an ISO 9000 system. The reference demonstrated ISO9001 briefly as follows:

> The standards themselves have been seen from two perspectives, the optimistic and pessimistic views. According to the former, the standards serve as the beginning of a total quality management (TQM) program. Issues such as internal organization, internal and external communication, and employee awareness of quality, product conformance, and customer satisfaction are all addressed, simplifying management commitment to quality. The pessimistic view holds that conformance to the ISO 9000 standards cannot be considered true commitment to quality on the part of management because the driving force underlying many registrations is the acquisition of the certificate itself and not the quality improvements that it brings. This perspective suggests that implementation of the standards may lead to excessive emphasis on the documented procedures and less emphasis on exceeding their requirements.

The message is that the “added value” that an organization derives from the ISO 9000 standards is a result of its true motives for, and approach to, implementation. In other words, are the standards adopted strictly as a customer requirement or are they a true quality initiative? Simple conformance to the ISO 9000 standards is not considered excellence; the success of a quality management program that builds on the foundation of the ISO 9000 system also relates to the original motivation for registration (Boys et al, 2003).

It is on a set of eight management principles that the guidance offered in ISO 9004 has been developed:

1 - Customer focus,
2 - Leadership,
3 - Involvement of people,
4 - Process approach,
5 - System approach to management,
6 - Continual improvement,
7 - Factual approach to decision making,
8 - Mutually beneficial supplier relationships.

They are considered to be the key factors in the ability of management to lead an organization towards improved performance. It is on the basis of these principles that the main ISO 9004 elements of management responsibility, resource management, product realization, and measurement, analysis and improvement were designed (ISO 9001:2008).

The role of people and documentation in support of the processes is highlighted, and the incorporation of results from verification and validation of processes as inputs into a process is stressed as a key to the attainment of continuous performance improvement.

2-3-4 1 ISO9001:2008 and the application of ISO9004

It is suggested here to present the standard ISO9000:2008, which has minor changes to ISO9001:2000, along with its pair ISO9004. While ISO9001 standard is a quality assurance standard by itself, it is more comprehensive and advanced to reach TQM concepts when applied with the concepts of ISO9004. This is also highlighted by Boys et al (2003), who mentioned that the ISO 9004 document was changed in both purpose and scope to become a guideline for performance improvement. The goal was to create ISO 9001 and ISO 9004 as a “consistent pair of standards”, so that companies that wanted to exceed the basic requirements of ISO

9001 could apply ISO 9004 to enhance their QMS and move towards business excellence.

The guidelines of ISO9004 contain the eight principles which the standard ISO9001 is based upon. It is useful here to present both.

Further, it is also emphasized that, although specific guidance is not provided within the document itself, function-specific management systems, including programs for environmental, health and safety, and risk management can be aligned or integrated with ISO 9004.

On the other hand, looking for ISO9001 as a certificate only will not be a TQM improvement technique; the reasons for this are explained by Leeman (2002). As quality award models are designed by business people for competition, such models are not based on every criterion that includes a quality dimension. Researchers have pointed out that even the winning of these awards doesn't necessarily lead to improvement in the quality of the operations assessed. A number of authors have highlighted the dangers of managers mistaking the implementation of TQM for quality accreditation. Critiques of quality awards are also well discussed in the literature (Leeman 2002), calling into question the ubiquity and completeness of these awards. Many of the award criteria identify the organizational areas and processes which need to be evaluated, but they do not stipulate how they should be managed in order to realize TQM. Ultimately these awards are intended to be devices for identifying good practices rather than formulating models for TQM.

Organizations which apply ISO9001, apply TQM fundamentals in an acceptable manner.

2 -3- 5 Malcolm Baldridge National Quality Award MBNQA

Some literature considers MB along with Deming and the Australian award to be the three major awards that are based on a perceived TQM model (Leonard and McAdam, 2002). Raisinghani et al (2005) added to the definition of MB: "the award is named for Mr. Malcolm Baldrige, who served as Secretary of Commerce from 1981 to 1987 and is credited for making sweeping changes that improved efficiency and effectiveness of government".

The model is based on 7 modules, derived from baldridge21.com (2/2010):

Leadership (120 points), divided into senior leadership and governance & social responsibility;

Measurement, analysis and knowledge management (90 points);

Strategic planning (85 points); workforce focus (85 points); Process management (85 points); Customer focus (85 points); Results (450 points), and further divided into Product outcome, customer focus outcome, financial and marketing outcome, workforce-focused outcome, process effectiveness outcome, and leadership outcome.

As can be seen, MB is similar to EFQM. EFQM has nine modules which are have the MB modules integrated inside.

The researcher's opinion is endorsed by the study of Leonard and McAdam (2002), who found that MB was considered to be the leading model until the EFQM adapted its key principles; in their view, MB also omits marketing and organizational design.

Although MB is considered an American quality award, and exclusively by its creators as a "national award", Leonard and McAdam (2002) consider MB as the most important catalyst in transforming American business, and that it, more than any other initiative, has reshaped management's thinking and behaviour.

Now if we consult other major national models derived from the Centre for Organizational Excellence Research, www.coer.org.nz, in the appendix, it is essential to mention that most of the models are based on EFQM or MBNQA.

2-3-6 Arabic quality awards

Referring to Labib (2006), it was found that literature relating to quality and excellence in the Arab world indicates that ISO certification for export to the EU is the driver for the adoption of formalized quality systems.

By exploring basic Arabic quality awards, it is found that some preliminary trials, such as the Dubai quality award and the Jordanian excellence award (for business association), are based on the EFQM model (Dubaided.gov.ae www.kaa.jo, www.kaaps.jo,www.associations award-jo.com, (4/2010)).

They are supported by the highest level of authority, Prince Bin Rashed in Dubai and King Abdullah Π in Jordan. It also has a different main element where there is large cultural diversity out there.

The Saudi standards organization (saso.org,sa 2/2010) promotes both MBNQA and EFQM along with ISO9000.

2-3-7 The Egyptian National Awards

The National Quality Institute NQI advocates this research and identifying, adopting, and selecting the best way of applying quality and productivity improvement in the Egyptian industrial and service organizations. (nqiegypt.org). NQI also adopted the Egyptian national awards for excellence.

This consists of many Criteria & Sub Criteria:

1.0 Leadership
1.1 Leadership Vision
1.2 Leadership Educational System
1.3 Leadership Support
1.4 Leadership Selection and Suitability
1.5 Creativity and Innovation
2.0 Strategic Planning
2.1 Strategic Development Process
2.2. Strategic Objectives
2.3 Action Plans
2.4 Performance measurement planning
3.0 Process Management
3.1 Product Process Management
3.2 Support Process Management
3.3 Supplier Processes Management
3.4 Customer Process Management
4.0 Resources Management
4.1.0 Human Resources Management
4.2.0 Knowledge Resources Management
4.3.0 Financial Resources Management
4.4.0 Material Resources Management
5.0 Business Results
5.1 Human Resources Results
5.2 Customer Satisfaction
5.3.0 Product and services Results
5.4.0 Supplier results
5.5.0 Financial and Market Results
5.6.0 Society Results

The Egyptian National Award has a comparative study in chapter four.

2-4 Quality Management Tools and techniques

This is the next level of quality entries; the tools and techniques. The following are the main tools, with more information to follow.

2-4-1 Lean Production/Manufacturing

Extending to Kaizen, the Lean concept needs to be defined here. Dollen et al (2008), who have carried out a literature survey, describe lean manufacturing as a philosophy focused on using less of everything – less material, less labour, less time, and less space than traditional mass production. There is evidence in the literature to suggest that lean manufacturing is associated with improvements in business performance measures.

Olivella et al (2007) reported that the term "lean production" is the "minimum workshop". As Ohno (the founder of Toyota Production system) says, LP can be achieved through various work organization schemes. For example, in the 1990s, Toyota changed from a seniority-based scheme to a skill-based one. In an LP plant, what happens on the floor is critical to solving problems and developing improvements. Therefore, line workers must have a high degree of influence, and together form a great body of knowledge. Manual workers become both learning and knowledge workers.

LP has some associated methodologies, such as just-in-time (JIT), total quality management (TQM), and total productive maintenance (TPM) (Katayama and Benett, 1996).

Katayama and Benett (1996), while exploring the role and significance of lean production within the context of the current industrial and economic environment, declared that a key feature is that fewer resource inputs are required by the manufacturing system (less material, fewer parts, and shorter production operations, less unproductive time needed for set-ups, etc.). At the same time there is pressure for higher output performance to be achieved (better quality, higher technical specifications, greater product variety, etc.). This should result in greater customer satisfaction which in turn provides the opportunity for the lean company to gain a market share larger than those of its competitors.

But the same authors added that a particular weakness of lean production is its inability to accommodate the variations or reductions in demand for finished products which have occurred in many companies. Only small changes in demand will often take production to below the break-even point. Therefore the new economic and competitive situation that is emerging calls for systems of production which can manufacture goods profitably across a range of levels of demand.

2-4-2 Kaizen

The term “kaizen” is taken from the Japanese words “kai,” meaning change, and “zen,” meaning good. Another definition of the Japanese meaning of kaizen is “to take apart and put back together in a better way”. The popular meaning is continual, incremental improvement of all aspects of a company (Doolen 2008).

Kaizen, as a Japanese tool for waste removal, has many features (Doolen 2008): their focus on low capital investment, their degree of action orientation and autonomy, their use of a cross-functional team structure, and their application of well-established quality and process analysis tools such as time observations, cause-and-effect diagrams, and other lean tools.

In this research, the same concept "low capital investment" is adopted from Kaizen, a concept that is most suitable for the EIO economically.

The intended impact of kaizen events is twofold: first, to substantially improve the performance of the targeted work area (e.g., a cell, production line, process, or even a product); and second to create positive human resource outcomes. Such outcomes are purported to create an organizational culture focused on longer-term continuous improvement. As Doolen (2008) mentioned, Kaizen is a process that takes about a week, while TQM is an approach that takes a long time. By using kaizen at multiple points in time, cycles of performance improvement within a given process are created.

Raisinghani et al (2005) give a fine description of Kaizen when saying that the best way to characterize Kaizen is to think in terms of process. The western management style focuses on results and is characterized as a result oriented approach; Kaizen is a process-oriented approach and focuses on small continuous improvements.

The literature advocates that Kaizen be promoted throughout the world. For example, Aoki (2007) mentioned that studies of kaizen activities in countries outside Japan suggest that the concept of kaizen has become routinely accepted throughout the world.

Aoki added that the Community of practice [of quality management approaches] involves a group of people who mutually engage in a practice through negotiating meaning, jointly pursue their own enterprises, and share resources such as routines, words, tools, and ways of doing things, and so on.

The proposed model uses the same tools mentioned above.

Also according to Aoki (2007), the relation of homogeneity between production, maintenance and engineering to analyse the root cause is as follows: in this process, communication between different functions has a critical role. If the cause of muda (waste elimination) has to do with machines and equipment, communication between the user of the machine, a maintenance person and an equipment (or production) engineer is essential to analyse the root causes. Once the results of such an analysis are reflected in the revision of a work standard, an improved standard is established as the basis for the next round of kaizen. In this process, communication between the people who make work standards and the people who are involved with the analysis is indispensable.

2-4-3 Design of experiment

Design of experiment is a systematic approach to investigation of a system or process. A series of structured tests are designed in which planned changes are made to the input variables of a process or system. The effects of these changes on a pre-defined output are then assessed.

Design of experiment can be used to find answers in situations such as "what is the main contributing factor to a problem?", "how well does the system/process perform in the presence of noise?", "what is the best configuration of factor values to minimize variation in a response?", etc. In general, these questions are given labels as particular types of study. In the examples given above, these are problem solving, parameter design, and robustness study. In each case, Design of experiment is used to find the answer; the only thing that marks them as different is which factors would be used in the experiment (thequalityportal.com).

2-4-4 Taguchi method

Taguchi methods, developed by Dr. Genichi Taguchi, refer to techniques of quality engineering that embody both statistical process control (SPC) and new quality related management techniques. Most of the attention and discussion on Taguchi methods has been focused on the statistical aspects of the procedure; it is the conceptual framework of a methodology for quality improvement and process robustness that needs to be emphasized. The entire concept can be described in two basic ideas (wtec.org)

1. 'Quality should be measured by the deviation from a specified target value, rather than by conformance to pre-set tolerance limits
2. Quality cannot be ensured through inspection and rework, but must be built in through the appropriate design of the process and product'

2-4-3 Just in time JIT

JIT is a time saving philosophy. Salaheldin (2005) introduced its requirements, such as modifying the material flow through many options (changing inventory and order policies, as well as production runs policies); reducing the number of vendors, stabilizing production schedules on a daily or weekly basis; planning production from final assembly; developing methods for estimating work-in-process and identifying why the company needs it, then trying to reduce it regularly; and establishing new procedures for dealing with suppliers, like defining the criteria for suppliers based on quality, cost and timing.

The implementation of JIT requires change in employees' performance, including training employees to improve their job skills in technical matters and problems solving; changing from individual to group incentives that are related to JIT accomplishment; reducing job classifications; substituting multi-skilled jobs; increasing flexibility of work reassignment; ensuring the ability to work overtime to complete daily schedules; participation of employees in decision making; training of management and employees to create an organizational culture consistent with the JIT philosophy; and building good relations with suppliers and improving communications between management and employees.

The research advocates that JIT is a concept that is hard to implement in Egypt because of the sharp time factor it requires.

2-4-4 Benchmarking

Benchmarking is seen as an important management tool of total quality management. The term benchmarking originated from the machine construction industry and refers to grouping technical and financial performance indicators for comparisons in units within organizations or amongst organizations (Magd, 2008).

Benchmarking has a specific way to go in order to get results. Results indicated that the most important reasons for initiating benchmarking are to maintain and increase competitive advantage, increase profitability, and achieve continuous improvement. Moreover, the most important benefits derived from benchmarking include improved customer satisfaction and improved response time. Further, top management commitment was found to be an important influential factor for effective benchmarking (Magd, 2008).

Benchmarking should be incorporated in the business planning process. As Magd 2008 noted clearly, benchmarking helps to identify the gaps that exist between current performance and best practice and also how best practice performance has been achieved, but in order for improvement to occur, a set of actions must be implemented. Regarding best practice, it is not necessary to identify the absolute best practice in the world in order for benchmarking to be successful. Good or superior practice is probably a more accurate phrase.

Benchmarking can be classified into four types: Internal, competitive, functional and generic benchmarking. For effective implementation of benchmarking, one should do the following: Be clear why a benchmarking exercise is relevant in the first place; Establish the type of benchmarking to be conducted (e.g. process and internal); Understand the nature of the business and its interactions with the environment; Select candidates and a specific benchmarking focus (e.g. process) that are representational of the whole company and are comparable (and to recognize the differences where not identical); Take a team-based approach and adhere to a project plan.

Benchmarking is considered as a tool of TQM, which can be used in organizations as a means of strengthening a company's ability to compete and achieve superior performance. Benchmarking has rapidly gained acceptance worldwide as a tool of continuous improvement.

In the researcher's view, benchmarking is one good quality management approach, but it is hard to implement in Egypt because of the scarcity of factual data across different entities.

2-4-5 Six Sigma

Coming from the United States, six sigma is a mathematical and statistical quality tool. Raisinghani et al (2005) give an elaborate briefing of six sigma:

> 'The introduction of Six Sigma into the manufacturing arena in the early 1980s by Motorola was a step in revolutionizing the scope and use of quality systems in business today. To define Six Sigma in simple terms is not possible because it encompasses the methodology of problem solving, and focuses on optimization and cultural change. Six Sigma accomplishes this goal by utilizing an extensive set of rigorous tools, uncompromising use of statistical and advanced mathematical tools, and a well-defined methodology that produces significant results quickly. The success of this

methodology within an organization has significant momentum that can only lead to fundamental organizational cultural transformation. The roots of sigma as a measurement standard go back to Carl Fredrick Gauss (1777-1855), who introduced the concept of the normal curve or distribution'.

The three sigma concept is related to a process yield of 99.973 per cent and represented a defect rate of 2,600 per million, which was adequate for most manufacturing organizations until the early 1980s. Two things occurred in the early 1980s that required a higher-level quality from American manufacturers. One of these was the introduction of mass produced miniature electronics, from transistor radios to televisions, which were produced in large quantities for mass-market consumption.

The second and more compelling force for domestic quality improvement was the opening of global markets and subsequent introduction of Japanese electronics into foreign and American markets. The lower price and higher quality of the Japanese goods made these imports attractive to the global consumer.

The quality level associated with a measure of Six Sigma corresponds to a failure rate of two parts per billion and it adopted this as a standard.

To illustrate why a 99 per cent quality level is not acceptable, consider the following facts:

- At major airports, 99 per cent quality means two unsafe plane landings/ day;
- In mail processing, 99 per cent quality means 16,000 pieces of lost mail/ hour;
- In power generation, 99 per cent quality will result in 7 hours of no electricity each month;
- In medical surgery, 99 per cent quality means 500 incorrect surgical operations per week;
- In water processing, 99 per cent quality means one hour of unsafe drinking water per month.

So, the immediate goal of Six Sigma is defect reduction. Reduced defects lead to yield improvement, and higher yields improve customer satisfaction. Six Sigma defect reduction is intended to lead to cost reduction. It has a process focus and aims to highlight process improvement opportunities through systematic measurement. Six Sigma implementation can have negative consequences if applied in the wrong project. Six Sigma is a toolset, not a management system and is best used in conjunction with other more comprehensive quality standards.

However, there are many other QM concepts or tools worldwide. Such a tool can be implied in one of the above, or it can be more technically generated in this context; examples include robust engineering, Quality function deployment (extracted from Aoki: SQC then TQC leading to QFD), and the Kanaban system.

2-5 Discussions of the demonstrated quality entries

There are many mile stones when demonstrating the many aspects presented and to emphasize the meaning.

2-5-1 Implementation of Business Process Reengineering BPR, Kaizen and TQM

Different names carry different meanings; BPR and TQM are contradicting concepts, and Kaizen and TQM vary greatly in timing. More implementation lessons are presented hereunder.

Doolen et al (2008), who carried out a literature survey which provides guidance on the most authority's readings on the subject, demonstrated that in contrast to BPR and kaizen, TQM typically focuses on an incremental process of improvement to improve customer focus and results through teamwork.

Although the targeted level of process improvement is different, the TQM literature stresses the need for employee involvement in driving and sustaining performance improvement. While many organizations appear to have achieved substantial process improvement through TQM, the literature suggests that many firms experience difficulty sustaining their TQM programs to drive long-term performance improvement. Researchers suggest such failures are due to poor program management.

The BPR and TQM literature both suggest that the process of improvement initiatives impacts the organization in two key subsystems: the technical system, which produces the core outputs of the organization, and the social system, which provides for the coordination of human resources and processes within the overall work system (Doolen 2008).

2-5-2 Similarities in the implementation of Business Process Reengineering BPR & Kaizen

Like BPR, kaizen promise dramatic improvements and uses cross-functional project teams. However, kaizen is designed to be low capital and have a short implementation cycle, while BPR often involves extensive (often organization-wide) investment in new information technology for enterprise business processes and requires more implementation time. Despite the promise of BPR, research suggests that the majority of these change efforts were not successful based on achievement of initial goals. However, Doolen (2008) found that the BPR efforts involving relevant employees and stakeholders more extensively using a team-based approach were more successful. These same organizations also reported higher employee morale.

Nevertheless, the researcher suggests that BPR, as a dramatic and shocking method, usually contradicts the normal "resistance to change" built into human behaviour. Consequently it is not welcomed by most employees.

2-5-3 Concepts of European Foundation of Quality Management EFQM & Balanced score Cards BSC

McAdam (1999), while relating EFQM to BSC, found that the Business Excellence Model BEM (EFQM) and BSC models attempt to link TQM and business goals in a number of ways. First, the nine criteria of the EFQM claim to represent a business and its strategy in its totality. Second, policy and strategy is a key criterion; and third the results' criteria are supposedly linked to the causal actions associated with the enablers.

Also, the BSC process seeks to link strategic statements down to business or continuous improvement activity within the operations of an organization.

Both the BEM and BSC specifically address the needs of customers and have sophisticated customer measuring systems. The BEM has a customer results criterion and the BSC has a quadrant dedicated to customers (McAdam 1999).

On the other hand, after reviewing the EFQM model thoroughly, Park (2008) found out that the best strategy for using the model as a management control model is "adaptation" rather than "adoption". This is the same conclusion that Park (2008) came to after having compared the EFQM's criteria with the critical success factors for innovation and new product development identified through a comprehensive literature review.

Table 2–1: Goals of quality standards with respect to management principles

<table>
<tr><th></th><th>Customers</th><th>Leadership</th><th>Objective</th><th>Personnel Development</th></tr>
<tr><td rowspan="3">ISO9001</td><td>Meeting customer requirements for products and services</td><td>Management responsibility for the quality effort</td><td>Consistency of action</td><td rowspan="3">Competence, awareness and training</td></tr>
<tr><td>Satisfying contractual obligations</td><td colspan="2">Measuring performance against established quality objectives</td></tr>
<tr><td>Measuring and analysing customer satisfaction</td><td>Demonstrating continual improvement</td><td></td></tr>
<tr><td rowspan="3">Six Sigma</td><td>Understanding customers' needs and action on them</td><td>Using data to drive decisions and guide the business</td><td>Reducing defects and variations</td><td rowspan="3">Building future leaders</td></tr>
<tr><td colspan="3">Achieving break through solutions</td></tr>
<tr><td colspan="3">Optimizing processes within the organization to reduce costs, improve productivity and increase customer satisfaction</td></tr>
<tr><td rowspan="3">Baldridge Criteria for excellence</td><td rowspan="3">Increasing competitiveness through customer driven excellence</td><td>Innovative and agile business management</td><td>Fact based approach to running the business</td><td rowspan="3">Placing value on employees and partners</td></tr>
<tr><td>Strategic thought and future direction</td><td>Benchmarking against industry best practice</td></tr>
<tr><td>Visionary leadership and public responsibility</td><td></td></tr>
</table>

To investigate adoptability of EFQM within management theories as an alternative management control model, Park (2008) found that the model can be adapted if its limitations are supplemented with other ideas or frameworks, such as BSC.

This advice is largely used by the researcher.

A considerable number of theoreticians and companies have made efforts in adopting and integrating the balanced scorecard model as a management control model (Park, 2008). The researcher is now one of them.

It is suggested here to borrow the definition of Management Control from the same author: Management Control is the process by which managers assure that resources are obtained and used effectively and efficiently in the accomplishment of the organization's objectives.

It can be seen that the definition is closely related to TQM concept.

It is true that some models are closer to the measuring concept, as McAdam (1999) mentioned, the business excellence model BEM (EFQM based) along with the Baldridge model, are the first systematic approaches for effectively measuring TQM within an organization. But the industrial gap in technology and human differences cannot be neglected while approaching other models.

For example, McAdam (1999) provides the fact that BEM & BSC require diversity of skills and experience across many different functions, so to tackle this point in the EQMA, there has to be human leverage, training on different soft skills and the devising of systematic experiences through succession plans, for example.

2-5-4: Goals of some entries in ISO9001, six sigma and Baldridge model

Table 2-1 will show how some tools and models have adopted some quality and management entries, Customers, Leadership, Objectives and Personnel Development, in different perspectives:

Table 2-2 shows goals of quality standards with respect to management principles, extracted from ASQ magazine, 3/2003, and summarizes some main dimensions of those models.

The main items taken from those standards into the posed model are: Consistency of action (From ISO9001), Building future leaders (From six sigma), Fact based approach to running the business, Strategic thought and future direction, and Visionary leadership and public responsibility (From Malcolm Baldridge).

2-5-5 Aspects of TQM concepts in some different quality entries

Louise (2008) presented a study comparing different quality management models (ISO9001, PQA, Deming, EFQM, MBNQA and Ford Q1) with respect to Total Quality concepts, and found that the ISO9000 scored the minimum grade; needless to say, ISO9000 standards are the dominant quality systems in Egypt so far.

The table below shows more results:

Table 2–2: Standards' aspects on Total Quality

Standards' aspects onTotal Quality	ISO 9001 :2008	PQA Philips Quality Award	Deming	EFQM	MBNQA	Ford Q1
Customer first	X	XXX	X	XXX	XXX	0
Leadership	X	XX	X	XXX	XXX	0
Value People	-	X	X	XXX	XXX	-
Customer/Supplier Relationship	0	XXX	XX	X	XXX	X
Excellence	-	X	X	XX	X	X
Policy Deployment	X	XXX	X	XXX	XXX	X
Process Management	XXX	XX	XX	XXX	XXX	X
Values	-	XX	XX	XXX	XXX	-
Business Chain	X	0	X	XXX	X	0
Bench Marking	-	X	0	XX	XXX	-
Policy & Strategy	XX	-	XXX	XX	XXX	-
People Satisfaction	-	0	X	XXX	XX	-
Impact on society	-	-	-	XX	0	-
Business Results	X	0	-	XX	X	-
Improvement process	XX	XXX		XX	XX	0
Total Score	**26**	**34**	**27**	**52**	**49**	**12**

- Not mentioned 0 minimal (1) X OK (2) XX very good (3)
XXX excellent (4),

In broad terms, the awards stress the importance of total quality, customer satisfaction, and people and process management to the attainment of superior competitive position. For example, the MBNQA focuses on leadership, and customer and market focused strategy. The

EFQM focuses on the relations with the community, and customers' as well as employees' satisfaction. The Australian Quality Award focuses on the interplay of organizational factors with the ends of customer focus and performance, (Leeman 2002).

Other standards like PQA or Ford Q1 were not further investigated for the purpose of this study as it only explores the most common quality management approaches worldwide.

The result of the literature review so far is that there is a gap between the philosophies of TQM and its application. It is the endeavour of the proposed model to cover this gap.

Also, there is a need to differentiate more deeply between various models presented; this is emphasized in the following chapter(s).

After Leonard and McAdam (2002) presented the combination and comparisons, they mentioned that the principles behind these approaches have been used for a number of years, during which time a methodology has been built in terms of the weightings and the criteria. These models are used extensively throughout the world and thus can be used for benchmarking. They have also helped TQM concepts to develop by defining tangible and intangible concepts and benefits of TQM.

Chapter Three

Comparative Studies of Quality Management Models

To structure the proposed HEQMA effectively, it is a good methodology here to develop a clear TQM vision integrating three different aspects of the journey to organizational excellence:

At the upper level there are the philosophies;
Then at the medium level there are the models;
At the lower there will be the tools and techniques.

It can then become the technical area of the sustainable excellence journey helping to find the right method needed for solving the research problem.

The proposed model is intended to cover all the components of TQM in EIO perception, in order to be advanced. The integrative and dynamic relations amongst the constituents of the model are also developed eventually throughout the research.

The table shows that HEQMA is using many of the TQM components. It is clear also that HEQMA is using the most dominant factors used by Deming and Juran.

3-1 Comparative Studies with quality gurus

Table 3–1: Comparative Studies and model integration with quality gurus

	Juran	Deming	Feigenbaum	Crosby	HEQMA
Management involvement in quality	√		√		√
Leadership		√		√	√
Quality planning	√	√	√	√	√
Quality control	√			√	√
Quality improvement	√	√			√
Customer orientation			√		√
Goals to all staff	√	√			√
Measurement	√				√
Recognition	√				√
Bench marking	√				×
Deploying training	√	√			√
Process orientation		√			√
No fear (Eliminate Slogans)	√	√			
Production of Variation		√			
Zero defects				√	×
Total 15 principles	**10**	**8**	**3**	**4**	**11**

HH3-2 TQM principles in a Quality Management model

As Park (2008) demonstrated, in the journey of QM modelling, Quality management (QM) is a management philosophy, which has evolved from a rather narrow and mechanistic approach known as statistical quality control, introduced by Shewhart to a more holistic and humanistic approach under the term total quality management (TQM) and business excellence.

During the last five decades the basic assumptions and paradigms of quality have constantly changed parallel with the general changes of paradigms in societies and global environments. These changes are not unique within the quality field but more or less in line with changes within other management fields including organizational theory and management control theory.

Despite the increasing focus on QM during the 1980s and 1990s, and despite the fact that quality became a central agenda for top managers, there have been relatively limited attempts on searching, reflecting and analysing its framework as seen from a broader theoretical perspective. In fact, this was also one of the major criticisms that the quality movement often received from various theoreticians. This criticism has been received in the quality research field with some surprise. One of the reasons is that the development of the QM philosophy and in particular TQM was based on an emergent need for a new management philosophy and a holistic/integrative management model, which had an emphasis on quality as the main strategy for improving competitiveness. The need for improving quality and thereby strengthening competitiveness was perceived as an emergent task for major European and North American companies, which were confronted with serious challenges from Japanese companies. The Japanese companies had become "masters" in practicing company-wide quality control (CWQC), and in this way they gained a competitive advantage compared to most Western companies. This company-wide approach to quality was unknown for most Western companies until the 1980s, when it was understood what had happened in Japan, and why a new management philosophy was needed. The huge interest toward TQM during the last part of the 1980s and the 1990s should be understood with these contextual factors as a background. The existing theories and models of organization and management could not provide the necessary principles, tools and systems to meet these new challenges and problems. For practitioners, the majority of the existing management theories were considered to be either too theoretical or too fragmental.

When viewing TQM in respect to some other management practices, it should be noted that TQM is a philosophical approach to management of an organization, and kaizen is a process that is not typically used by managers who subscribe to TQM (Emiliani 2005).

As TQM recommends applying a system of quality standard, this view is compatible with the researcher's, where the quality systems come as a complementary item in the formulated model HEQMA. All models are originated and based on TQM concepts.

It is suggested to check the critical success factors for a similar developing country as India. It is found that the following factors are the crucial ones:

1. Organizational commitment
2. Human resources management
3. Supplier integration
4. Quality policy
5. Product design
6. Role of quality department
7. Quality data reporting
8. Technology utilization
9. Operating procedures
10. Training

(El-Araby and Abdel Ghany, 2007).

While developing the new measurable approach, amongst many international models, it is inevitable to mention a similar effort that summarizes what the research is working with. McAdam (1999) formulated the five TQM pillars which this research used as criteria for selecting the work items. He mentioned that his model, derived from the qualitative data, lists the key critical factors under the five category headings, when implementing quality models. Many of the factors listed apply equally to large and small organizations.

TQM embodies comprehensive approaches or philosophies, such as those previously discussed, towards continuous improvement and the elimination of waste from the production process. In addition, TQM places a great deal of importance on issues regarding the manufacturing infrastructure such as workforce development and training, managerial leadership and commitment, team-based work organization, and the use of small groups or teams to solve problems and develop new production methods, most of the TQM components mentioned here are included in HEQMA.

TQM has also become a very popular area of research in recent years. Several authors have proposed models out of TQM, most of these models are based on theories and practices that are primarily derived from the manufacturing industry (Leeman, 2002).

The TQM-performance link represents another major topic of academic research in recent years. If customers are demanding improved quality, then what QM approach leads to the best organizational performance? What managerial practices produce the highest quality and customer satisfaction?

The proposed approach tries to answer such questions.

Commitment to quality is one of the measures that seem to be controversial. It is very beneficial within the context of this study to define it in elaborator manner.

Lee & Peccei (2006) define the term Quality Commitment in their study as the degree to which the top manager shows identification with, involvement in, and responsibility for performing quality jobs.

3-3 Business Process Reengineering (BPR) in a model

Tennant and Wu (2005) have found that the main reasons for applying the BPR technique were external competitive pressures, internal cost reduction, and productivity improvements. According to the same authors, the main barriers to successful implementation were reported as tactical short-term solution-driven approaches, which inadequately considered people issues, and an over reliance on IT-based technology.

The researcher considers the above two reasons are even more valid in Egypt, and this is obvious in many public sector factories which were sold to the private sector and investors who restructured the organizations, a main tool of BPR, and dispensed of man-power, thus causing many social problems and strikes.

3-4 Comparative studies with major models

While the Japanese model is similar to EQMA in the influence of top management commitment, EFQM and MBNQA are still the dominant models that lead and inspire other nations' models, as per the results of the Centre for Organizational Excellence Research, and the attached relevant appendix. So, here they will be compared with EQMA.

3-4-1- European Foundation of Quality Management EFQM and HEQMA

There are 23 interactive elements within the 2 models:

Table 3- 2: Comparison between European Foundation of Quality Management EFQM and HEQMA

	Top Mgmt. Commit	Corp. Ethics	Tr.& Devt.	Team Perf.	Employee Motive	Improve Productivity	COQ	Safety & Society	Mktg. Results	CSI
Leadership	√			√	√					
People Mgt.		√		√	√					
Strategy	√									
Partnership &resource			√	√	√					
Processes, prod.& serve						√	√	√		
People results		√			√					
Customer results									√	√
Society results				√				√		
Key results						√	√		√	√

It was also mentioned that there is updated compatibility between HEQMA and EFQM in many ways.

If we go back to explain the methodology of EFQM, Lluster et al (2005) do this by linking enablers and results as follows:

> 'The EFQM Model constitutes a non-prescriptive framework that assumes there are different approaches to achieving sustainable excellence that derives in the existence of multiple interpretations around its implementation.
>
> However, it is made up of certain notions and ideas about the general relationships between its elements that have still not been demonstrated empirically. In this sense, the logic behind the model is that by improving how the organization operates; the "enablers" – there will be an inevitable improvement in the "results".
>
> The study of the relationships between the elements of the EFQM Excellence Model must take into account of two main issues. Firstly, how the enabler and result criteria are they interrelated and to what extent an integrated approach should be adopted. Secondly, the model presupposes an association between the enablers and the results, since excellence in the enablers will be visible in the results. Consequently, the causal relationships within and between both domains emerge as important aspects to be analyzed'

This same rationale underlies several studies that empirically test a number of causal relationships between some of these elements in the EFQM Excellence Model, as in other quality awards such as the Malcolm Baldrige National Quality Award; however, it is widely recognized that the empirical research in the causal relationships within the EFQM Excellence Model is still limited, since it is mostly based on studies that test isolated associations.

As was presented earlier, Park (2008) checked the disadvantages or limitations of EFQM, as it is almost obvious that a simple model will not be able to cope with the complexity of a system with a high degree of uncertainty and unpredictability.

Now the model is discussed regarding whether to adopt it and how to adopt it in relation to Egyptian culture. Park (2008) added:

> 'The culture aspect in terms of value, vision and mission building was explicitly focused under Leadership, while this focus was more or less ignored in policy and strategy, partnership and resources as well as in the process criterion. These inconsistencies seem to be a major defect of the model and may have been the reason for many companies' problems with implementing the model as an overall framework for strategic planning and improvement of the business'

However, the concepts of self-assessment and scoring the elements are still valid for HEQMA.

On reviewing EFQM changes that took place in 2009, the researcher found it useful not to indulge in the details of the EFQM indicators so as to reduce distraction and bias. However, there are many cultural factors which cannot be just imported from Europe to the Egyptian industry.

On the other hand, EFQM is sponsored in Egypt through the Egyptian union for excellence. In the interview with them, they declared that the advantage of EFQM is its key factors and sub-criteria; it is direct; it has the RADAR approach. The hard part in implementation is to sustain results. They also expressed that we need to force the role model intensively.

Finally; they proposed the creation of an EFQM customized to the Egyptian culture; this could be material for future research.

3-4-2 Malcolm Baldridge National Quality Award MBNQA and HEQMA

Comparing to the MBNQA, there are 36 interactive elements within the two models as follow:

Table No. 3-3: Comparing HEQMA to MBNQA

	Top Mgmt. Commit	Corp. Ethics	Tr.& Devt.	Team Perf.	Employee Motive	Improve Prody.	CoQ	Safe& Society	Mktg. Results	CSI
Leadership (Sr., Gov. & social resp.)	√	√			√			√	√	
Strategic Planning	√						√			
Customer focus						√	√		√	√
Workforce focus			√	√	√					
Pss. Mgmt			√			√	√			
Measure, ana&Know. mgt	√		√			√	√			
Product outcome						√	√			
Customer outcome									√	√
Finl.&mktg. Outcome						√	√		√	
Pss. effectiv Outcome						√	√	√		
Leadership Outcome	√	√		√	√					√

If it is considered that the elements are integrative and can then be in a line proportion, the estimation within the two models can be expressed as follows.

As EFQM has 9 main elements, while MBNQA has 12, then the relative interactive elements of EFQM are about 31 elements. Still, MBNQA has more interactive elements with HEQMA.

On the other hand, cultural nominations are clear that EFQM considers "people, results" while MBNQA considers "workforce, outcome" for the same meaning.

3-5 Comparative study of implementing quality tools

In the following paragraphs, a link is demonstrated regarding how major techniques can or cannot be successfully implemented in Egypt:

3-5-1 Six Sigma with respect to HEQMA

Six Sigma accomplishes its goal by utilizing an extensive set of rigorous tools, uncompromising use of statistical and advanced mathematical tools, and a well-defined methodology that produces significant results quickly. These techniques are hard to apply in an environment with high uncertainty like the Egyptian industrial climate which also faces increased economic pressures.

The success of the six sigma methodology within an organization has significant momentum that can only lead to fundamental organizational cultural transformation, Raisinghani et al (2005):

Again the cultural factor in Egyptian organization cannot stand such major transformation.

The quality level is another factor for looking at six sigma with suspicion. For example, in the Motorola way for improvement, it was found that the quality level associated with a measure of Six Sigma corresponds to a failure rate of two parts per billion and it adopted this as a standard.

This caused six sigma to be coined as a trade mark for Motorola, where it is considered that a 99% quality level is not acceptable.

Moreover, six sigma depends on many deep quality tools, which are not used in Egypt, such as Measurement system analysis, Process control, Design of experiments, DMAIC approach (Define, Measure, Analyse, Improve and Control), Failure Mode and Effects Analysis (FMEA), and Quality control and capability analysis.

Six sigma can be seen, due to the above, as a sophisticated subject in the Egyptian business environment where the lack of accurate databases or information may be apparent in many cases of business organization.

3-5-2 KAIZEN and Toyota Production System TPS

For Kaizen and TPS, a good situation in Egyptian industrial companies would be if they can reach the situation described by Ihara (2003, cited by Aoki, 2007) of the shop floor conditions in Toyota, which are as follows: the prefabricated building is enclosed by glass walls on all sides, the inside of the building and toilets are always kept clean, and even the placement of boxes is done with extreme care. He states that, in this situation, every worker always feels that he/she is under someone's supervision, and thus it is mentally easier for the worker to work as he/she is instructed.

This situation makes it possible to discipline workers to conform to the company's rules and ways. This suggests that organizational capabilities which discipline workers to conform to the company's ways also play a certain role in Japanese kaizen activities.

3-6 Comparative studies with other researchers' models

When comparing the proposed model with the Sinclair and Zairi model (The European Centre for TQM, Bradford University), which was based on effective process management through performance measurement, the model, as a total quality-based model, includes five sections:

- Strategy development and goal deployment; this dimension is covered in the Top management commitment and corporate ethics bundle of the proposed model.
- Process management and measurement; this section is clear in the internal capability bundle of the model, but with emphasis on the industrial culture needs in Egypt, which is Q.C.
- Performance appraisal and management; this is typical in the HR capital bundle of the model, also customized to Egyptian parameters of team work.
- Break-point performance assessment which are efficiency-driven; and
- Reward and recognition systems section; a dimension that is new to enforce in the Egyptian culture through syndicates and labour unions, lately expressed by many strikes; this section is inherent in the

proposed model through income-competence and job security plans indicators.

Each section of the Zairi model is introduced and validated by the results of a survey of the performance measurement systems in a postal survey of 115 companies.

Looking also at the proposed items in the model of Walden and Gop (1996; mentioned in Leeman, 2002), they have the items as follows:

Top Management commitment / *Quality Culture* / Quality Data and Reporting / Quality Design / *Training / Process management / Customer satisfaction.*

It is found that the five items in italics are identical in both models; Quality Culture item is reflected here as Corporate ethics (Values, Attitudes and Behaviours).

in another comparison of the current proposed model with the model of Wilson and Collier (2000; mentioned in Leeman, 2002), it has the main factors of:

Leadership (represented here as top management commitment), process management (Internal Capabilities), HR management (reflected as employee motivation, training and education), Strategic Planning (as part of top management commitment), Information and analysis (part of Cost of Quality analysis), Customer focus and satisfaction (the same in the proposed model).

Moreover, the criteria for selecting the HEQMA elements include:

- To embody business management, not only quality management. This is why there are detailed elements on marketing, market share, and export.
- To embody all stakeholders. This is why employees should benefit from the approach as well as the owners, and the surrounding region where the organization is located should benefit as well as the whole society.

In some of the other literature, Bossink et al (1992) developed a TQM model which contained 8 elements:

Totality (Quality management to be concerned with an organization as a whole) - Line-staff relationship - Technological perspective - Cultural implantation - Management commitment - Upstream emphasis - Market-in approach - Integration.

Some of the elements mentioned above are similar in nature to the main HEQMA elements. They are cross-interpreted but in detailed indicators of HEQMA, for example Line-staff relationship is detailed in Team performance and employee motivation, Cultural implantation is detailed in corporate ethics, Management commitment is the same element, and the Market-in approach is detailed in the quality marketing item. The significance in the current study is in giving a measurable dimension to enforce the meaning.

The Bossink et al (1992) model also has four application areas:

1- Primary activities; i.e. the operational activities;
2- Supporting activities such as personnel management and manufacturing control;
3- Conceptual/ regulatory activities, for instance: product/market combinations and allocation of resources;
4- Top management activities.

Again, the four application areas cover the items illustrated in HEQMA as follows:

- Item No. 1 above is interpreted in HEQMA process capability.
- Item No. 2 above is interpreted in HEQMA human capital.
- Item No. 3 above is interpreted in HEQMA marketing and corporate ethics.
- Item No. 4 above is interpreted in HEQMA top management commitment.

The above literature review describes what is known so far in the TQM arena. Surely, some TQM conceptual models are pioneering, but lack empirical validation. Contrarily, some well-designed empirical studies tend to lack theoretical support, especially those award-based frameworks.

The above literature review also highlights what remains unclear and under-examination. It also suggests topics needing further research and validation.

First and foremost, the journey of theory development should continue and focus on specifying and testing the relationships among key TQM variables. Secondly, given the inconclusive findings regarding the impact of TQM on profitability, there is a pending need to develop other effective performance measures, such as cost of quality. Lastly, in order to avoid the bias of using managerial perceptual data as a surrogate of customer satisfaction, the incorporation of an actual satisfaction index should be seriously considered. As Leeman (2002) expressed:

> 'Given the differences in conceptualization and the discrepancies of findings, empirical assessment of a comprehensive TQM model is called upon. This proposed model should provide a framework for launching an overall quality improvement program in organizations, showing the impact of both strategic and operational factors on product quality'

3-7 Formulating the proposed HEQMA from TQM principles

Thousands of businesses have embraced TQM principles in order to improve their competitiveness. Thus it is considered the most popular set of business principles over the past two decades. TQM was first utilized by Japanese manufacturers but has been promoted in the USA as a means of gaining a competitive edge.

From the above extensive research, a TQM model is developed in HEQMA. This model describes the basic conceptual elements of TQM. It also provides the way in which the basic elements can be made operational in Egyptian practice. Based on this developed approach, its items are tested in focus groups/interviews. The results will hopefully look promising for purposes of using the indicators for realizing TQM and 'measuring' and stimulating continuous quality improvement in EIO.

For example, to link production and marketing as two integrative items (in HEQMA), Pekgun et al (2006) illustrated that, in many firms, manufacturing is evaluated as a cost centre that seeks lower costs and operational efficiency, while marketing is evaluated as a revenue centre with control over price and other marketing elements. However, this is not necessarily an effective strategy. Dividing a firm into independent units for measuring performance on accounting terms would lead to misaligned incentives and suboptimal system performance. There is a need to emphasize the alignment of manufacturing and marketing incentives with the firm's goals and objectives. It is empirically demonstrated that business performance is enhanced when manufacturing and marketing work together for goal attainment.

On the other hand, proving the need for HEQMA, some questions are raised about the feasibility, source and need of current models. Conti (1999) demonstrated that the ISO 9000 standard is landing in a rather complex and somewhat critical scenario, made even more complex by a revival of the two best known TQM models: EFQM and MBNQA.

In fact, the Malcolm Baldrige Award underwent substantial changes in 1997 and the EFQM has launched the Business Excellence Model, gaining in competitive zest, but losing its European flavour.

A question arises and it seems to be justified in the light of recent history: Does such frantic activity in such models reflect the actual needs of users? Or does it rather meet the need for competition among 'suppliers', who measure their success according to how widely their model is used? Are we going to see competition among a number of models, one of them an international standard and the other two *de facto* standards, which try to cut out their captive markets based on a geopolitical range of influence? Would such an evolution benefit or damage the economy and quality of life?

The above reference provides more evidence regarding the need for developing HEQMA, when he declared that:

The general trend today is to move upwards from quality assurance to TQM, to organizational excellence models. Both standards and award organizations want to keep or improve their positions as leaders in quality management. Therefore, all of them claim that their models cover the widest possible area, even the most advanced areas. It is the firm belief that only correct positioning of the different models could avoid distortions that often cause damage to economic development. Both international quality standards and award models have their institutional areas to cover. Clearly defining those areas and focusing on them is an important goal.

Conti (1999) added that in a free world and a free economy, progress is closely connected with the intensity of debate and with comparison, differentiation, variety of ideas and experience. Quality has been an important issue in the debate and research of the past two decades. However, in the areas where qualitative differentiation could still generate significant improvements in the relationship between generated value/resources employed, the introduction of standards can curb development.

The focus should be on rationalizing supply relationships and breaking down technical barriers that are preventing the free movement of goods, as well as on dealing with people's safety and protection of the environment, where the reasons for introducing standards are self-evident.

The above mentioned items are included in HEQMA.

So when coming to develop new approaches, Conti (1999) cleared the way to this trend by stating that models intended to improve the competitiveness of specific companies, in particular self-assessment, should be left to evolve by themselves. A major authority 'proclaiming' an official model should not restrain free evolution.

The present study is an explicit expression of this trend.

On the other hand, if the application of TQM has many obstacles, amongst them being management non-commitments and performance measures being unavailable, the HEQMA model provides indicators for such measures. The model provides for practical applications of TQM concepts, which may not be gathered in other models, such as Cost of Quality, corporate ethics, and values measure.

3-7-1 Guideline criterion for establishing HEQMA

The following rules are derived from TQM fundamentals of formulating a model, they are inculcated in HEQMA:

(1) TQM is strategically linked to the business goals:
Approaches must cope with rapidly changing environments and be adaptable;
The focus must be wider than simply financial (in our case, it must be wider than quality as well);
(2) Customer understanding and satisfaction are vital:
A wider range of products and services can be developed; Improved target setting for markets and customers;
Improved organizational alignment with customer needs;
(3) Employee participation and understanding at all levels are required:
There is an increased focus on training and development; there is an increased emphasis on helping employees learn, innovate and improve; Employees are recognized as a source of innovation; Appropriate measures are developed for learning and growth; Reward and recognition for employee efforts; The need for careful communication with employees;
(4) The need for management commitment and consistency of purpose:
The strong central focus of management aids implementation; Management commitment is vital; learning to create future value as well as short term gain;
Managers must avoid using HEQMA as a tool against employees; Long implementation times are a test of management resolve; Managers must communicate regularly and effectively;
(5) The importance of processes and measures:
There is a danger of developing too many measures; There is a constant tension between the need for flexibility and the constraints of processes and measures; Many managers prefer "doing" rather than "measuring";

Considerable scarce resources are required to capture measures on an on-going basis; Training and development are needed for effective process management;
Targets can be established for processes; Process benchmarking can help overcome the parochial nature of management processes (McAdam 1999).

It is intended here to diagnose the need of TQM in Egypt and what items specifically may be needed for the Egyptian environment.

Here, systemic approaches for achieving sustainable excellence are discussed, taken from EFQM guidelines. It is important to realize that only a systemic, systematic, and integrated approach, like those embedded in the previously mentioned models and frameworks, is capable of steering the right way to reach an outstanding performance model.

Systemic: There are three major types of system definitions:

An ensemble one: a system is an ensemble of elements and relations in dynamic interactions.

An instrumentalist one: a system is an organized whole of processes related by an interactions network.

And an energetic one: a system is a set of transformations, which in a general way, by the play of these interactions, are controlled, while managing flows of energy.

All these frameworks integrate a holistic view of organizations and of organizational performance:

Systematic: Systematic means that these frameworks require a total and generalized involvement. No one in the organization can escape from them once it is decided to adopt them. To support such as exceptional involvement, an awards system can be used. It will help to create challenge and urgency, hope and combativeness.

Integrated: Few people involved in Total Quality Management have a clear vision of integration and efficiency of their own function. Many ask if it is better to adopt ISO9001 or an excellence framework, or the six-sigma and Balanced Scorecard methods. In some organizations, Quality assurance and TQM belong to different departments. TQM is subordinate to business objectives like all other process of the organization.

Adopting an excellence framework and coupling it with generalized RADAR is a good practical way to link the strategic alignment of the organization with that of its maturity.

The final question is which excellence model is best to choose?

Using the excellence model of the geographical area because of the cultural differences as mentioned beforehand. Such a community of interest is very helpful to experiment with an excellence model and to compare the experience with other people having tried the same thing with other opportunities.

It is the researcher's role to convert such literature into a contribution. The literature review supported the pilot study with respect to the research problem integration.

The literature review proved the gap between the philosophies and application. HEQMA endeavours to fill this gap.

By the end of both chapters 2 and 3, it is concluded that international quality models used to be in a quality assurance phase, when ISO9000 used to be the wedge which prevent the cycle of plan-do-check-act from deteriorating downward.

Nowadays, quality management models have become ‘radarized’ (i.e. are using the RADAR approach). Both new models are revolving in the frame of process management.

This is what EIO should realize and ascend to after treating the subject of "missed Q.C.".

CHAPTER FOUR

IMPLEMENTING QUALITY MODELS IN EGYPT

In this chapter, we follow a journey to explore what models are implemented in Egypt, what the lessons learnt are, and then back to the future to use the historical background as a base for improvement of the proposed model.

4-1 Quality Management Models implemented in Egypt

The intention is to highlight specific case studies of applying quality entries and their implications in Egypt.

4-1-1 Balanced Score Cards BSC

A useful starting point is provided by Ismail (2007) who carried out a survey on 150 big Egyptian companies. In his research results, the main obstacles of implementing a quality model in Egypt were the lack of management information systems and/or information, along with management attitude. This will draw our attention to concentrate on those factors when formulating an Egyptian quality management approach.

It is a good selection of the author that companies are listed in the Egyptian stock exchange market, but to find that 60% of them are implementing BSC is something to be questioned; that model doesn't seem to be so popular in Egypt, it doesn't have enough credible consultants to be able to upgrade such a number of organizations.

In the findings, there seem to be some contradictions where it is said that the BSC has wide spread use in the Egyptian companies surveyed, but the level of use of multi-dimensional indicators is significantly low (Ismail 2007).

It can be seen that BSC is based on multi-dimensional indicators.

Nevertheless, the study proves that there is success for that model, which invites the current research to customize some of the bright edges of the model.

While BSC can be suitable for private sector companies who have the flexibility to deal with many parameters and to change accordingly, the survival of the organization depends on its ability to both evaluate current performance and identify strategies to improve the quality of planning and control decisions. To help them do this, company managers have used a variety of performance evaluation tools and measurement techniques. Many studies examine management accounting practices and performance evaluation measures, along with individual benchmarking dimensions, in different developed countries.

However, there is a lack of empirical research examining Egyptian private sector practices of using performance evaluation measures and analysing factors that may affect the selection of those measures.

Some surveys conclude that many companies have found the BSC to be a useful tool for focusing and sustaining their continuous improvement efforts. Research indicates a company's strategy may also be important when investigating a company's choice of performance measures (Ismail 2007).

So, for the BSC, the researcher found it more suitable to advocate its customized presentation to the Egyptian industrial society. Here, managers can focus on both long-term and short-term measures of success. Of course the shortfalls mentioned in chapter two had to be taken into consideration.

It is seen that for the success of BSC, it is necessary to adapt it as a whole in the Egyptian business domain, as organization structures in Egypt commonly don't have multiple layers that the BSC usually works best with.

4-1-2 Just In Time JIT

In introducing the implementation of JIT in Egypt, Salaheldin (2005) claims that:

> 'JIT philosophy can be successfully implemented in Egyptian manufacturing companies; just like their peers in developed countries including Japan, USA and the Far East; with thoughtful attention applied to each individual company. He added also that; Egyptian manufacturing companies can use JIT as a strategic philosophy, rather than as a temporary solution for operations obstacles'

In another view to the mentioned reference, the researcher suggests that JIT actually depends, to a large degree, on suppliers' performance and strict supply of long-term agreements, which can barely be achieved in Egyptian business organizations.

Based on the survey result, Salaheldin (2005) claimed that about 60% of his survey Egyptian companies were implementing JIT. He indicated in the findings that about 40.4 percent of companies participating in the empirical study had not implemented the JIT philosophy. They reported that limited knowledge about JIT is the biggest obstacle impeding its implementation.

Contrary to the above result, logic and facts reject that almost 40% of the organizations don't know about JIT, while the other 60 % are suddenly applying JIT, with no grading of probability in between.

It is assumed that the percentage of JIT-implementers in Egypt is much lower.

Egypt, like most less developed countries, strives to diagnose and find solutions to the severe problems that are obstructing the growth and development of its industrial sectors. Problems like high scrap, losing market shares, high levels of inventory, poor quality in products and labour, long lead times and the existence of many sources of waste in production processes. It is controversial that JIT can help to remove those problems.

As Salaheldin 2005 discovered:

> 'Research has shown that the successful implementation of JIT philosophy can produce significant benefits for manufacturing firms; such as, improving quality that consistently and continually meets customers' requirements; minimizing levels of inventory and improving relationship with suppliers; reducing the labor turn over rate; reducing manufacturing lead times; reducing set-up time; reducing operations and materials handling costs; and maximizing the use of space.
>
> JIT can also improve the on-time receipt of material from suppliers; improving purchasing function; improving preventive maintenance; increasing worker participation; improving the quality and timing of received material; full utilization of people, equipment, materials and parts; and improving competition while reducing paper work.
>
> Contrasted with this is the fact that caution is needed because the arrival of materials must be accurate and continuous; plus the use of backup inventories is necessary.
>
> Moreover, it is suggested that the effectiveness of JIT implementation hinges on a wide range of special support that includes: high quality supplies; strong management commitment; a manageable supplier network; geographic concentration; plus efficient transportation; and materials handling'.

In discussion of JIT implementation in developing nations, Salaheldin (2005) also concluded that:

> 'For JIT practices in Mexico; JIT can be used successfully in some Mexican manufacturing firms.
>
> Furthermore, it is founded that there are three groups of obstacles that hinder the implementation of JIT in Mexican operations:
>
> (1) Employee participation obstacles.
>
> (2) Supplier participation obstacles.
>
> (3) Obstacles to the managerial integration of the JIT companies'.

The same reference added that, in a study conducted on the implementation of just-in-time in manufacturing firms in Ghana, they found that JIT firms differ from non-JIT firms in terms of their efforts with regard to set-up time reduction, continuous quality improvements, suppliers' partnerships, and employees' training. However, there is no significant difference with regard to the use of measurement systems.

Then the author adds, for JIT modifications, researchers and practitioners indicated that several modifications to existing systems should be undertaken prior to JIT implementation:

First, JIT requires a modified approach by top management which may include significant modifications, such as designing an organization that integrates strategy with people to achieve the basic premise of JIT, elimination of all types of waste; reducing specialization and organization functions; creating project teams; making everyone responsible for production of quality products and services; developing management and employees' commitment to continuous improvement; and integrating people and organizational systems with hardware.

Second, the engineering modifications of JIT include: changing the work centre layout; combining several operations to minimize the distance travelled; grouping machines in cells; purchasing equipment with short setup; responsibility for product design; quality and reliability; using design of experiments to improve quality while reducing costs; seeking product standardization wherever feasible; concentrating on continuous improvement in product design; using total productive maintenance (TPM) as an integral part of a JIT system; plus analysing the operations in order to identify where standardization, simplifications and automation are needed.

Third, JIT requires modifying the material flow through changing inventory and order policies, as well as production runs policies; reducing the number of vendors; stabilizing production schedules on a daily or weekly basis; planning production from final assembly; developing methods for estimating work-in-process and identifying why the company needs it, then trying to reduce it regularly; and establishing new procedures for dealing with suppliers, like defining the criteria for suppliers based on quality, cost and timing.

Finally, the implementation of JIT requires some human resources modification efforts, including training employees to improve their job skills in technical matters and problems solving; changing from individual to group incentives that are related to JIT accomplishment; reducing job classifications; substituting multi-skilled jobs; increasing flexibility of work reassignment; ensuring the ability to work overtime to complete daily schedules; participation of employees in decision making; training of management and employees to create an organizational culture consistent with the JIT philosophy; and building good relations with suppliers and improving communications between management and employees.

The factors of JIT success in Japan need strong efforts to attain in Egypt. The control over suppliers and the lack of good information systems are the main obstacles for JIT implementation in Egypt.

4-1-3 Bench Marking

Benchmarking has been gaining attention among managers as a means of strengthening a company's ability to compete and achieve superior performance. It has also rapidly gained acceptance worldwide as a tool of continuous improvement.

Magd (2008), in his study of understanding benchmarking in Egyptian organizations, declared that there is a clear spread of benchmarking across the various industry sectors in Egypt and this suggests that benchmarking is viewed as an important tool for the future in those organizations. The researcher examined the relationship between firm sizes and the presence of undertaking benchmarking activities. The results suggest that a higher proportion of the medium and large organizations were performing benchmarking activities than small ones, and large organizations are more likely to be benchmarking than small firms. This is probably because of the lack of time, resources and knowledge in small firms. However, small firms showed some interest in benchmarking, since 14 of them were using it. Therefore, it may be said that organizational size has no direct impact on the utilization of benchmarking activities or not.

However, the most important influential factor in Egyptian organizations was top management commitment. This finding is not surprising, as it is consistent with other studies.

The same reference reported that the problems with benchmarking in Egypt were that 60 percent of his surveyed sample was not involved in benchmarking. The main reasons for not benchmarking are data comparability and lack of resources (time and money). In addition, some of the respondents (17 percent) felt that their current company practices were sufficient.

Benchmarking is a useful tool for continuous improvement based on objective measures and concise data, however, the researcher considers that benchmarking is not suitable for the Egyptian business climate as information disclosure, which is the basis for benchmarking, will not be successful amongst different organizations working in the same field. This is natural in the closed and competitive market of Egypt.

Also, benchmarking needs continuous data and statistical analysis, this is barely used in most Egyptian industrial operations.

So, for benchmarking, little empirical research has been carried out in the Arab world, and more specifically Egypt (Magd, 2008).

The same study suggests that Egyptian managers should be aware of the importance of top management commitment, employee involvement and participation, and effective communication for effective benchmarking implementation. He found that 60 percent of the surveyed sample was NOT involved in bench marking. The main reasons for this are data comparability, and lack of resources (time and money).

4-1-4 World-Class Manufacturing WCM

When comparing Egyptian manufacturers to WCM practices, Salaheldin and Eid (2007) found in their study that Egyptian manufacturers are still in the 1970s and 1980s, when compared with world-class manufacturers. More importantly, the results of this study indicate that poor planning and lack of knowledge are the most significant barriers to WCM implementation in the Egyptian manufacturing sector.

This situation may be factual, as the WCM is a more advanced level of industry. Egyptian manufacturing may need to fulfil some QMS standards and concepts first in order to be compatible with WCM. It is hoped to create more awareness among management and employees about the strategic importance of WCM techniques to operations processes in the Egyptian manufacturing firms.

It can be concluded that non-awareness of new practices, such as e-commerce, Electronic Data Interchange EDI, lean production, Enterprise Resource planning ERP, and supply chain management, can guarantee that Egyptian manufacturers will fall apart after World Trade Organization

(WTO) ascension. In other words, it will keep them too far behind any competitive advantage that can be acquired in the manufacturing world today.

In their study on the implementation of WCM in Egyptian manufacturing firms, Salaheldin and Eid (2007) followed a procedure to analyse the implementation of excellent manufacturing techniques by Egyptian manufacturing firms, the researchers developed two major questions:

(1) What progress have Egyptian manufacturers made towards World-Class status and where are they currently?
(2) What are the drivers that encourage Egyptian manufacturers to adopt these WCM strategies? And what are the difficulties which may prevent them from embracing such strategies?

These were the overall questions to be answered by their current study, which were defined by the following three objectives:

(1) Explore the current status of Excellent Manufacturing Techniques applications and management practices in Egyptian manufacturing;
(2) Identify the critical driving and resisting forces toward WCM techniques implementation in Egyptian manufacturing firms; and
(3) Provide guidelines for the successful implementation of WCM by Egyptian manufacturers.

The results of this study indicate that the Egyptian manufacturers are still using the mass production philosophy. The most modern practices to emerge in the 1990s were considered as the least performed by the Egyptian manufacturers which means that they are still dominated by the old style "mass production theme" and have not yet been touched by the new world class theme of "mass customizations."

Considering the WCM drivers and barriers, the results of this study are consistent with previous studies that see that WCM implementation is driven by the never-ending needs of customers to look for better services and products. However, the need to avoid losing market share to competitors is a key incentive, since this research has found that competitors' use of the WCM and response to customers also has a strong (driving) effect on the adoption of the WCM techniques.

At the same time, the need to improve the "range and quality of services" relates to IT developments are forcing organizations to be up-to-date in their use of advanced technologies regarding the production and delivery of speedy and high quality information, as well as facilitating greater degrees of communication and integration across business units and external partners.

To sum up, the Egyptian manufacturing firms are driven to adopt the WCM techniques because of the incentives or gains achievable through the use of these new techniques where advantages outweigh disadvantages regardless of the barriers that might decrease their use.

The researcher agrees with the above findings.

4-1-5- ISO9001

In a valuable piece of research, Magd (2006) collected some data for implementation in 70 Egyptian companies of ISO9001:2000, and he found that the participants perceived top management commitment and the lack of qualified personnel to be major barriers for the effective implementation of ISO 9001:2000. Again this will point the current research to the importance of measuring top management commitment and to the qualifying of the work force through training and development.

The same reference claims a relation between product quality and system quality. It is vital for the companies to gain certification because more and more European and foreign buyers have become frustrated having to verify the quality of Egyptian goods they purchase; a costly and time consuming process (Magd 2006). Meanwhile, gaining certification to ISO9001 or any other model proves only that the company has the ABILITY to produce quality products.

The same mixed concept came from respondents of the same study, where the researcher stated that it is clear Egyptian manufacturing companies considered internal improvement of the organization’s PRODUCTS and processes and external pressure to obtain the certificate as the main driving forces for seeking certification (Magd 2006).

When the volume of documentation (documented procedures) is reduced by 67% from ISO9001:1994 to ISO9001:2000 (ASQ magazine, March 2003), it is shocking to discover that improving the documentation and the efficiency of the quality system appear to be the leading benefits for implementing ISO 9001:2000 in Egyptian manufacturing organizations (Magd 2006). This finding means that Egyptian organizations didn't benefit from the improvement applied to the ISO9001:2000 version.

On the other hand, according to the same reference, the most important barrier was perceived as top management commitment, and this was due to the fact that without top management commitment, employees tend to put a low priority on the QMS and it becomes static.

Moreover, if top management does not show commitment to the importance of the system continuously, employees will prioritized other work activities over the quality management system, and nothing will result in terms of the effective implementation of the quality management system (Magd 2006). The previous statement abbreviate many situations faced by consultants (the present researcher used to be one of them) where top management of many organizations tend to degrade the theme of quality management with respect to other business parameters.

Again it is obvious that top management dedication to quality deployment should constitute a significant part of the proposed HEQMA.

It seems astonishing, as well, to find a similar study by the same reference is giving a false indication that is far from the previous study.

As Magd and Curry (2003) claim, the results of their study indicated that Egyptian organizations have a high level of understanding of the purpose of ISO certification.

On reviewing the state of ISO9001 certification in Egypt, Salaheldin had demonstrated in his findings, the majority of ISO certified companies pointed out that adopting ISO certification means implementing TQM (Salaheldin 2003). Therefore the researcher suggests that Egyptian companies still need much training and awareness on basic quality concepts.

The author faced a problem while conducting his data collection. He studied some companies that do not apply TQM, or maybe they do not yet know it clearly, while he is researching with them the driving and inhibiting forces for TQM implementation. The research also did not define for the questioned companies the entries of TQM to measure them against.

In connection to the above, it is found that top management commitment is the strongest driving force for TQM implementation in the Egyptian industrial sector (Salaheldin 2003). The same study structured the restricting forces that inhibit the TQM strategy in Egypt as insufficient infrastructure, lack of training, workers reluctance to get involved in decision making, and inadequate knowledge base.

The results indicated above are considered sufficient for the purpose of this research, taking into account the lack of academic research in this area. Magd (2006) also found that little research on the quality management model has been carried out in developing countries, and more specifically Egypt.

4-2 Historical background of quality approach in Egypt

Egypt, as a part of Africa, is characterized by some cultural domains. As Labib (2006) indicated, African countries have a high power distance (refusal to change) compared to USA. Power differences in an African hierarchy tend to be greater and more accepted, thus paternalistic and authoritarian management styles are more common.

For Egypt, in the past there was the Egyptian standardization and quality control authority, developed to be the Egyptian organization for standardization and quality EOS. This organization has promoted quality culture and getting companies certified to quality marks, for products, and quality system standards, mainly the ISO9000 family, for management systems. In that time, many consulting offices entered the market to upgrade the companies' systems to be adapted to the ISO requirements.

In this phase, the consultation activity took place and flourished to fulfil company needs for certification. In the next section, a discussion is given on the implementation of ISO 9001 in the Egyptian industry as a whole.

The role of consultation was endorsed by some granting agencies, such as centre of quality assurance CQA, an affiliation to US Aid, then industrial modernization centre IMC, affiliated to EU partnership with the Egyptian government.

Recently, EOS has promoted the Egyptian National Award for Excellence.

4 -2-1 The Egyptian National Awards

National Quality Institute (nqiegypt.org), NQI adopted the Egyptian national awards for excellence, which consists of many criteria and sub-criteria mentioned previously:

1.0 Leadership
2.0 Strategic Planning
3.0 Process Management
4.0 Resources Management
5.0 Business Results

As can be seen, the approach has taken many items out of MBNQA, and most importantly, the model mentioned many of the TQM components without saying how to apply them or what their dimensions are. This leaves the items in large controversial issues.

The Egyptian National Award for Excellence is a collection of merit awards initiated by the Industrial modernization centre IMC to value the industrial companies, and has three types. One of the recent awards is the national award for excellence in quality (nationalawards-eg.com). The ministry of industry and trade also delegated the process of the award to the IMC (imc-eg.org 2/2010) which supports consultations, training and certification through grants. Some employees are then nominated to assess and certify the awarded company on the concepts of leadership, strategic planning and efficient use of resources through specific procedures. However, the researcher's experience showed that the process of awarding and related information seem to be restricted to IMC.

In the review of the said award, the first section of its five is of no comment as it includes all the ingredients the top management commitment needs. The second section, strategic planning, contains a good new dimension, which is performance measurement planning. Planning in EIO is a necessity, and this is why it is inculcated in HEQMA.

For the three remaining model sections, below is a comparison of them:

Table 4-1 Results

	HR results	Customer Satisfaction	Product& Service	Supplier results	Financial& market results	Society results
Product Process Mgmt			√ Joined			
Support Process Mgmt						
Supplier Process Mgmt				√		
Customer Process Mgmt		√				Society has no action item
HR resource Mgmt	√					
knowledge resource Mgmt						
Financial resource Mgmt					√	
Material resource Mgmt				Material has no result item		

For knowledge resource Management: it has no result item
For financial resource Management: Market has no action item

The review of the current quality models inspired the researcher with more ideas to crystallize the proposed model; however, other inputs for the model are derived from the researcher's practices of management and quality concepts inculcated during the years.

This will result in determining more specifically the items from those models to fit Egyptian industrial and business circumstances, and thus will lead to focusing on the practices and indicators that formulate and extract EQMA.

4-3 General obstacles of implementing QM models in Egypt

As a part of the focus group interviews, the following interview results are mostly about the factors that hinder the good application of a model in Egypt. They are demonstrated here as expressed by the interviewees:

1 – There is a demand in the first place to explain the differences between EIO and other cultures (for example, Asian culture is contradictory to Western management, but both are successful).
2 – The problem with ISO9001 is that it was implemented by jumping over quality factors, such as quality control.
3 – The standard ISO 9001 is obligatory; it is connected to export facilitation and exemptions.
4 – Management commitment to TQM is weak in Egypt; training will also be enforced through public relations. The current industrial climate doesn't allow the application of a QM model.
5 – PEST becomes very influential when business grows, up to a certain volume.
6 – HEQMA is not a substitute to a certain model; as we only apply the appearance not the core.
7 – Education and culture greatly affect the corporate ethics; the normal worker level doesn't respond to the group value system.
8 – We need strong and honest leaders who will make the customer satisfied; as with European reputation contradictory to Chinese products, for example.
9 – Even the value system is via top management judgments; they would judge sometimes by emotion, not real performance measures. Also leadership influence depends on job level of the employee.
10 – To succeed, HEQMA has to be implemented gradually.
11 – The shortcoming is in people when implementing the value system.

12 – HEQMA success conditions are:

- Avoiding politics
- Offering the authority to the group implementing HEQMA (the society itself is an obstacle for implementing HEQMA).

13 – There are four challenges for success:

- The pressure for change
- Sharing a clear vision
- Allocating capability and resources
- Then comes HEQMA as the model; The First action.

It can be concluded that the reasons for failure to implement a QM model in some organizations are:

- Changing cultures and concepts, which takes long time.
- Resistance to change; inherent inside human beings.
- Implementation of results takes a long time; sustainability factor.
- Mis-implementation.
- Insufficient infrastructure, lack of training, workers reluctance to get involved in decision making, and inadequate knowledge base, are regarded as restricting forces that inhibit the introduction of HEQMA to EIO.

4-3-1 Draw backs of ISO9000 in the Egyptian Industry

The way to apply ISO9001 was one of the reasons that urged the researcher to tackle this research topic. As it is required to initiate total and planned changes in the concepts, styles and approaches of the Egyptian management, being satisfied only with achieving an ISO9001 quality management award is no longer valid. It seems apparent in this context that the ISO9000, as the dominant QMS implemented in Egypt, was cascaded by force of governmental orders to the public sector companies, while private sector companies were promised export customs deductions and other exemptions for ISO9000 holders.

Applying ISO9001 or other QMS forms by themselves didn't help Egyptian Industrial Organizations' EIO to achieve such high performance results.

Despite the large number of ISO9000 certifications in Egypt, which are over 1500 certificates in the past 15 years (www.eos.org.eg), Egyptian industries are still considered less developed. Some other companies cautiously tried some other quality management models, such as Balanced Score Cards and six sigma.

Nevertheless, money, effort, and time have been spent just on importing those models. The gap still exists between EIO and their counterparts in other developed countries.

Moreover, the implementation of ISO9000 Quality Management System (QMS) has only been done formally. This is endorsed by the literature (Salah Eldien, 2003) which found that activities related to quality are often misunderstood.

Awareness and education in Total Quality Management (TQM) in the industrial society seems to be poor. While reviewing TQM status in Egypt, it was found that the majority of ISO companies pointed out that adopting ISO certification means implementing TQM, whereas the ISO certification should be regarded as the first step towards TQM. The researcher's view is that the first steps towards TQM are creating values and attitudes within the organization, then establishing the system.

Magd and Curry (2003) admit in their study that there are few studies that have investigated management attitudes towards ISO9001:2000 in Egypt. On a regular basis, quality is equated with the checking activity of quality control, or understood in terms of product grade or excellence. Quality management systems are not new, but the introduction of QMS as new vocabulary might lead to confusion over its purpose and the way it is understood.

The following table, which is extracted from the Quality progress magazine of ASQ (March 2003), illustrates the change of QMS focus in ISO9001: 2000:

Table 4-2

Concept	Number of times mentioned in ISO 9001:1994	Number of times mentioned in ISO 9001:2000	Percentage of change
Improvement	0	20	+∞
Customer satisfaction	1	7	+600
Top management	3	11	+233
Documented procedures	18	6	-67
Processes	26	41	58

It seems that the EIO didn't cope with the new changes, apart from the slight changes that took place in the year 2008, such as enforcing outsourcing and integration with Environmental Management System ISO14001:2004.

Since ISO9000 is the most commonly deployed model in Egypt, in this respect, a Manchester University survey revealed that more than one third (approximately 40 to 50%) of the organizations implementing ISO9000 didn't benefit from the subjects of Employees awareness & motivation – Productivity improvement – waste reduction – marketing improvement – cost cuts – gaining new markets – export increase (Louise, 1999).

4-4 Imposing HEQMA to fit for Egyptian Industry requirements

The study of Maddi (1998), cited in Almukaadem, Mustafa (2009), provided integrative description for future plans of organizations in Egypt. The study revealed that TQM in the 21st century in the Egyptian organizations needs to adopt customer orientation (detailed in item 4-1 of HEQMA), process focus (detailed in item 3-1 of HEQMA), Fact based decisions (detailed in item 1-1 of HEQMA), continual improvement (detailed in item 3 of HEQMA), teamwork and spirit (detailed in item 2-2 and 2-3 of HEQMA), shared vision between management and work force (detailed in item 1-2 of HEQMA), and quality improvement (detailed in item 3-2 of HEQMA).

As can be seen, the imposed items for Egyptian organizations are refurbished in the texture of HEQMA.

There was exploration of many American approaches, BSC, 6 sigma, MBNQA; European ones, such as EFQM; Japanese ones, such as JIT, Lean, and Kaizen; and Integrative ones, such as Benchmark, WCM, TQM, ISO, and BPR.

After all, the proposed HEQMA takes from many of these:

- Tools and techniques; as in lean and Kaizen in the process capability item.
- Quality philosophies; TQM and Crosby in Cost of quality, training and education, safety and society protection, teamwork and motivation, and ethics.
- Not only quality models: BSC; details of some sub-items. EFQM and ISO9004; self-assessment. Scoring system as for EFQM and MBNQA.

This can provide its originality and value: a measure for performance improvement MPI.

CHAPTER FIVE

ASSESSING HEQMA VALIDITY AND FOCUS GROUP

The objectives of voting are to apply culturally-oriented indicators of HEQMA to be accountable-for-measure and to be able to benefit from versatile industrial and academic-quality backgrounds in the validation process. Such a model matches the Egyptian culture, then.

Items of MPI are integrative. However, excellence cannot be achieved unless ALL items are achieved.

5-1 Reasons for selecting focus group method

1 - Because the ultimate goal is to get deep information direct from the interviewees. Designing a postal questionnaire would not be suitable for this type of research as it would not get certain information off people's chests.

2 - By demonstrating HEQMA elements, taking different comments about its importance, role, modification, and practical examples, a focus group helps to relate items of the model to each other.

3 - The flexibility and content enriching nature of face to face communication is one big factor of interview success; it requires much experience to control the conversation. By concentrating on specific points, it helps greatly in the consequent analysis.

4 - A pitfall of interviews is the tendency to be biased (either by the researcher or the interviewees) feelings and behaviours. But the realization of problematic dimensions, with sincere trials for fairness, helps the researcher a great deal.

5 - Content analysis aims to reach the right reasoning for the data. Considering its context, it is suitable and very beneficial to have the open questions and comments.

6 - The method of interviews does not work properly when the resulting data is to be entered into deep statistical analysis (Dti.gov.uk).

7 - The nature of the present research is qualitative evaluation, not quantitative analysis. Its validity is achieved through experts and systematic scientific reviews.

8 - It is noted that the focus group is not applicable to the formula of the number of questionnaires related to the number of variables. For interviews, around 10 are sufficient for the research; this is different to the sample size used in mail-questionnaires (El-Kott, 2008). The pilot study used 9 interviews and 17 interviews for the focus group itself.

5-2 Voting on HEQMA elements

As this approach was accountable for measurement, interviewees reviewed the items and gave their scoring.

1- Base factor: %
1-1 Top management commitment
1-2 Corporate Ethics
2- Human capital factor%
2-1 Training and development
2-2 Team performance
2-3 Employee motivation
3- Internal Capability Factor %**:**
3-1 Improving Productivity
3-2 Cost of Quality
3-3 Safety and positive impact on society
4- Outcome factor%**:**
4-1 Marketing results and competitiveness
4-2 Customer Satisfaction Index, CSI
Comment:

5-3 Procedures of validation and focus group

The focus group was pretested and revised through the pilot study on a sample of academics and non-academic specialists in performance evaluation, as detailed in chapter one. Reliability and validity was done through the SPSS program.

Nevertheless, validity is gradually built up initially by the pilot study, and then by the literature reviews of quality gurus, international models, and other researchers' models. TQM concepts were also checked to finally be validated by the focus group technique.

The method of asking the interviewees and then using their answers to determine what attributes are provided in the model, makes it a good one.

Reliability and validity of the interviews was checked by ensuring the following: respondents are asked about topics on which they can respond; questions are appropriate; explaining why the data and information are needed make the request for information seem legitimate and increase the respondent's willingness to answer; and finally sensitive topics were tackled smoothly in the discussion, the same as Labib (2006) indicated.

By doing this, validity is naturally based on the chapters' construction/sequence, the accumulated output, and each part is tied to the following till the end.

Referring to validity literature, Mussallam (2007) mentioned that validity is the exact meaning of what is intended to be measured; it has many courses:

1 - Faith validity: This is where the first impression of interviews and feelings gives a view of what is done.
2 – Content validity: This is when the method covers all the topics of the subject.
3 – Construct validity: This type of validity was conducted in the present study with two interviewees. It measures consistency of different question (element) selections, where a continuously odd answer may reflect that something is wrong. It is like having grades of 4A then C in a subject, indicating non-consistent results.
4 – Empirical validity: This measures method related to the target of the process.
5 – Concurrent validity: This is where the scoring of items (selection tools) and consequent performance are matched to each other.
6 – Predictive validity: This is using scores given to predict performance (like the GMAT test).
7 – External validity: This is to generalize the effect(s) of the analysis.

The focus group included clarifying the HEQMA objectives; procedures included interviews and discussions with the experts and academic scholars on the following topics, as well as voting on them and giving comments:

1 - Viewing the implementation of foreign quality models such as ISO9001, how can HEQMA be implemented successfully as a national approach in Egyptian industrial companies?
2 - How will external factors, Political, Economic, Social, and Technological, affect HEQMA implementation?
3 - How will the training and development, vocational and technical, influence reaching HEQMA success?
4 - Do you see the Egyptian industrial climate as more suitable for "team leadership" rather than a charismatic inspiring leader?
5 - Compared to a successful corporate value systems (obeyed by and benefiting ALL staff, applied in justice and dignity), can an Egyptian Corporate Value System be established and positively affect the motivation, performance improvement and training?
6 - Will the implementation of HEQMA help Egyptian Organizations' top management in achieving customer satisfaction and successful marketing?
The following scale is used to respond to the questions:

Do not agree/NA	Fully Agree			Barely Agree
0	4	3	2	1

5-3-1 Proposed Influence Matrix (Action-Results matrix)

This method tests the strength relation(s) between the initiators that influence specific results; the relations marked here are only a proposal by the researcher. In the focus group itself, relations are kept blank to get the opinions of different interviewees on the model:

From the proposed relations, it is noted that the results needing the most action and effort are "Corporate Values" and "Team performance". This is natural, as values are greatly influenced by many parameters.

The action that leads to many results is "Training and development" because training and development are most influential when taken seriously.

Table 5-1 Actions-Results relation of HEQMA elements

Actions	Results: Results of Cost cuts and waste reductions	Corporate Values	Team performance	Cost of Quality	Marketing results and competitiveness	Customer satisfaction
Top Mgmt commit		√				√
Attitudes and behaviours		√	√			
Safety	√			√		
Impact on society		√	√			
Motivation			√			
Training & development		√	√	√	√	
Improving Productivity	√			√		
Customer focus					√	√

5-3-2 Company details

This panel of data is to check the characteristic features of the focus group:

1 - Company operating for: less than 5 years - 5 to 20 years - 21 to 35 years - more than 40 years.
Having considerable business experience within their sector, business experience is positively related to a company's ability to develop performance evaluation measures related to their strategic objectives.
2 - Number of people employed: less than 100 employees – 250 to 100 people – 500 to 250 people - 1000 to 500 people - more than 1000 people.
This is correlated with a company's need to maintain a general policy and objectives of performance evaluation that encourages employees to refine different process complexities, reduce costs, and encourage continuous learning.
3 - Computer information systems (CIS): organization filing system is based on separate files - Traditional manual data entry systems - CIS based on database systems - knowledge management.
4 – Company marketing strategies: Cost leadership - Differentiation - Focus strategies.
5 – Company Performance measures include: Financial measures (profit margin etc.) – Customer satisfaction – Market share – Defect percentage – Others: Please specify.

5–3–3 The selected focus groups included

1 - The academics were from the Arab Academy for Science and Technology and Maritime Transport and from the production engineering department, Alexandria University; one of the interviewees was an observer member in the ISO organization, while the other one was an EFQM representative in Egypt.
2- Industrial experts were from the following companies:
- El-Araby home appliances company (Quality general manager);
- El-Araby for lighting Technology Company (this company has achieved the EFQM award, level 1: Commitment To Excellence CTE, and is working on level 2: Recognition For Excellence RFE), (Quality manager);
- Alexandria Petroleum Maintenance Co. Petromaint; chairman;
- Amria pharmaceutical Co., QA team;

- Pharco, QA dept.;
- Pirelli Tires Co., production and quality managers and production section head (M.Sc.);
- Egyptian Petrochemical Co., department manager.

5-4 Focus group Conclusion

The throughput includes the following:

1 - Creating consistency between HEQMA elements by eliminating similarities, grouping the elements, and illustrating the elements that are needed for the stakeholders and customers;
2 - Grading the elements to reflect importance;
3 - Summing up HEQMA indicators in their numbers and length;
4 - Inserting the focus group recommendations and remarks such as:
- Adding the training needs assessment (TNA) to the indicators,
- Measuring authority-responsibility equity,
- Changing the term "Education" to "Development",
- Considering the relative weights, and constant comparison of items, shading the blind areas of action-results matrix, if any,
- In the item "improving productivity", an important indicator of man-hr productivity is added,
- Reducing the 360 degree feedback as an assessment system, as it can collide with cultural barriers in the superior-subordinate groups,
- Removing indicator(s) which do not relate to the item heading or the context,
- Interviewees recommended simplifying some expressions and reducing the number of technical indicators.

5-5 -Relative weights of HEQMA elements

It is proposed by the researcher that the weights are related to variable settings. The base factor (independent) and the outcome factor (dependent) are the dominant factors of the HEQMA, thus having higher relative weights (30 points each). The other two factors (moderate and intervening) represent the catalysts of the approach (each has 20 points).

Data analysis includes multiple readings through various channels, such as interviewing a focus group, transcripts, field notes, and documents given to the researcher by the interviewed experts. Interviewees were asked to explain tangible examples in detail in order to improve the reliability of the data.

5-6 Adopting quality control into the model and the Cost of Quality "COQ"

As indicated earlier, quality control representing the quality of products has been bypassed in many Egyptian manufacturing companies; therefore it has been focused on in the model. Improving the productivity item and sub-items reflects this idea. According to Doolen (2000), the focus should be on improving existing processes, rather than implementing solutions that require investment in new technology.

Cost of Quality, COQ, is also considered one of the most forgotten or underestimated dimensions of current quality models. This view is endorsed by Schiffauerova and Thomson (2005) who found that COQ is not utilized in most quality management programs. According to them, Quality guru Crosby (1983) states that nothing in his 30 years of work as a quality professional has disappointed him as much as the way the concept of COQ is not used.

Fortunately they also found that companies that do adopt COQ methods are successful in reducing quality costs and improving quality for their customers. Schiffauerova and Thomson also pointed out that most examples confirm that quality improvement and cost measurement processes bring about a huge reduction in a company's COQ.

It must be mentioned, according to the same reference, that the selected COQ model must suit the situation, the environment, the purpose and the needs of the company in order to have a chance of becoming a successful systematic tool in a quality management program.

Types of quality costs are:

1: Preventive Costs: Planning of business processes.
- Supplier selection and Evaluation;
-Training costs;
- Implementing a quality management system (QMS);
- Using statistical methods for forecasting and targeting, including cross-functional teams to control data;
- Production data analysis for preventing errors.

2: Appraisal costs: All Inspections and Tests, including laboratories, manpower and materials, indoor & outdoor Inspections and Tests.

3: Failure costs involve: Repairs, complaints, warranty, corrective actions, manpower and material waste, failure maintenance, scrap, failure of system (procedures etc.), Production data analysis for correction of errors.

Also, elements of improving productivity can include the following: Lead-time, set-up time, part travel time, percentage of on-time delivery, defect rate, floor space, throughput, and product design measures such as price, product line diversity, etc.

From the above, it is demonstrated here as well that each type of industry will propose and customize specific COQ items suitable for that industry. As Schiffauerova and Thomson also realized, a realistic estimate of COQ and improvement benefits, which is the trade-off between the level of conformance and non-conformance costs, should be considered an essential element of any quality initiative. Therefore, the elements included in COQ models of various companies differ substantially.

However, any recommended list of COQ can serve only as a guideline and for provoking thoughts. A suggested list is particularly useful if it is by-industry, because each particular industrial sector has its unique quality cost elements.

5-7 Training & development in the model

As a developing culture in the EIO, KSA (Knowledge, Attitudes & Skills) should be adopted and measured. Therefore it is highlighted in HEQMA.

Doolen (2008) provided useful definitions on the subject:

> 'Knowledge refers to the "body of information necessary" for the employee to perform tasks; skill refers to the "Psychomotor abilities" needed to perform tasks with "ease and precision;" and attitude refers to the "cognitive capabilities" required to perform the required tasks (e.g., desire to perform the given activity)'

The same reference suggests that active employee participation in the change process is likely to generate both higher levels of business performance improvement, and favourable employee attitudes.

On Job Training (OJT) is vital for the expertise succession through generations in such cultures as Japan. Aoki, in his study to designate the importance of OJT, suggests that Japanese companies that emphasize the importance of tacit knowledge have developed organizational capabilities that facilitate knowledge creation. In this perspective, the key agent of knowledge creation is a group of people who have various experience and knowledge.

OJT plays a critical role in creating such capabilities. Employees in Japanese companies experience various kinds of jobs through OJT, which helps to reduce social distance between different categories of the workforce. In this perspective, it is organizational capabilities, which facilitate communication among diverse people, and that allow Japanese companies to implement incremental organization-wide innovation (Aoki 2007).

It is worth mentioning here the power distance dimension of the Egyptian culture and how such OJT can help in this regard.

CHAPTER SIX

BALANCING THE RESULTS AND HEQMA

Early signs from the focus group results indicate that HEQMA can help to produce a measurable success in the Egyptian Industrial Organizations EIO. Another achievement of HEQMA development would be addressing interests of all stakeholders, namely Investors, Customers, Top management, Employees, Suppliers, and society. The routine quality approach's focus is ultimately the end users or the buying customer.

6-1 Data analysis results

6-1-1 Relative weights of HEQMA items

In general, the Base factor now has a score of 27 points (initially this was 30 points).

The Human factor now has 29 points (initially 20 points) as team performance has the biggest effect/recommendations regarding the Action-results matrix, and the most required principle is training.

Internal capability had a slight change (initially this was 20 points), while the outcome factor was regimented (initially 30 points) for the sake of the true stakeholders: internal customers; people.

Top management commitment had a consensus of votes, as is shown in the next figure; it took the highest overall average in voters. This indicates that this item is crucial to HEQMA success from the point of view of EIO.

Table 6-1

	Mgmt Commit.	Ethics	Train& Devt.	Team Perf.	Motive	Improving Productivity	COQ.	Safety, Society	Mktg results	CSI
General Av.	*13.9*	12.2	12.1	8.5	8.1	8.35	*7.6*	8.5	10.5	10.6
Acad. Av.	12.75	11.1	11.75	8.25	7.75	10.5	8.5	7.75	10	11.5
Indl. Av	14.2	12.5	12.15	8.7	8.15	7.7	7.4	8.77	10.6	10.4
Median	1.45	1.4	*0.4*	0.55	*0.4*	*2.8*	1.1	1.02	0.6	1.1
Res. Scoring	14	13	12	11	6	7	7	8	11	11

In general, the Base factor now has a score of 27 points (initially this was 30 points).

The Human factor now has 29 points (initially 20 points) as team performance has the biggest effect/recommendations regarding the Action-results matrix, and the most required principle is training.

Internal capability had a slight change (initially this was 20 points), while the outcome factor was regimented (initially 30 points) for the sake of the true stakeholders: internal customers; people.

Top management commitment had a consensus of votes, as is shown in the next figure; it took the highest overall average in voters. This indicates that this item is crucial to HEQMA success from the point of view of EIO.

Consensus on Management commitment item:

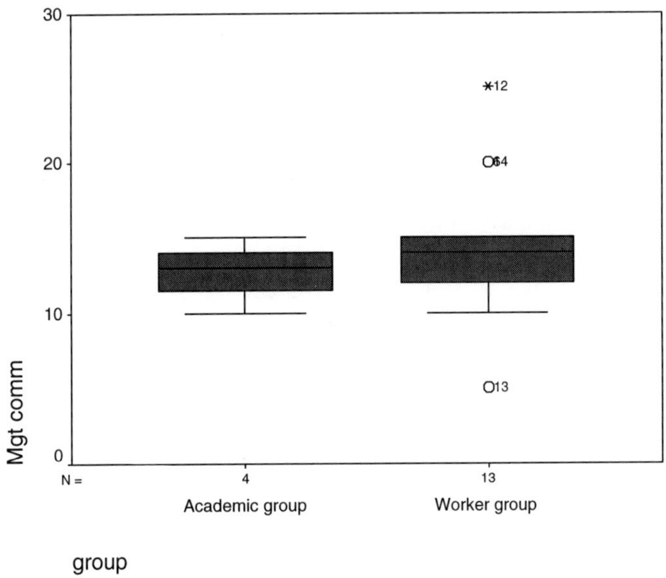

The minimum general average is for the "cost of quality" COQ item, identical to the industrial average. This is justified by the fact that counting the cost of quality is a new concept to the EIO and can be of relatively high cost and be hard to attain.

The biggest difference between academics and industrials was for the "improving productivity" item, which academics ranked as a higher value to consider. However, points of view between academics and industrials were identical for the "training and development" item. This indicates that the two groups vary when it comes to the technical domain, while they are on the same lines regarding the human factor domain.

6-1-2 Answers and comments of focus group

Table 6–2

	Q1: HEQMA as a model	Q2: PEST effect	Q3: Tr.& Devt.	Q4: Team work	Q5: Values	Q6: model input to output
General Av	2.35	2.6	*3.2*	2	2.12	2.29
Acad Av	2	3.25	3.75	3	1.5	2.75
Indl Av	2.5	2.4	3.1	1.7	2.3	2.15
Median	- 0.5	0.85	0.65	*1.3*	- 0.8	0.6

For topic 1, Feasibility of HEQMA compared to foreign approaches in Egypt, about 60% (2.35/4) expected success to be average; an encouraging response.

For topic 2, External factors (PEST) will largely affect HEQMA implementation; the difference is not big enough between academics and industrials. Academics assign high influence to PEST (Political, Economic, Social and Technological) factors, while the industrial experts lessened the ratings for these factors. This shows that those inside the system have more ability or confidence to receive or make a success of the factors of new management approaches; the researcher hopes that PEST effects are positive anyway.

For topic 3, although the training and development of people had the highest average of HEQMA items, after management commitment, some serious comments were provided that the cultural domain (regarding training) would adversely affect that factor.

For topic 4, cultural character, leadership specifically has the greatest significance between academics and industrials. Whether EIO should adopt group teamwork or concentrate on individual participation and distinction, most of the industrial experts tend to consider a "charismatic industrial leader" as very influential, while on the contrary, the academics encouraged "team leadership".

This dispersion of thoughts reflects the current cultural status of the Egyptian society and the industrial society in the middle of it, stimulating other researchers to tackle such crucial subjects separately. However, the researcher thinks that the level of education and culture of employees will affect the suitable options; this is endorsed by some focus group opinions.

For topic 5, comments reveal that the society itself may be a hindrance in front of the Egyptian corporate value system. Meanwhile, employee motivation would benefit a great deal from such a value system.

For topic 6, the final one, the chance that HEQMA will build itself up on a national level has an average of 57% (2.29/4). Again the comments put a burden on the external factors and the climate, rather than the top management's will. Some interviewees also interrelated heavily between topics 1 and 6, noting that they need more deep investigation.

The answers to the six questions were as follows:

Q1

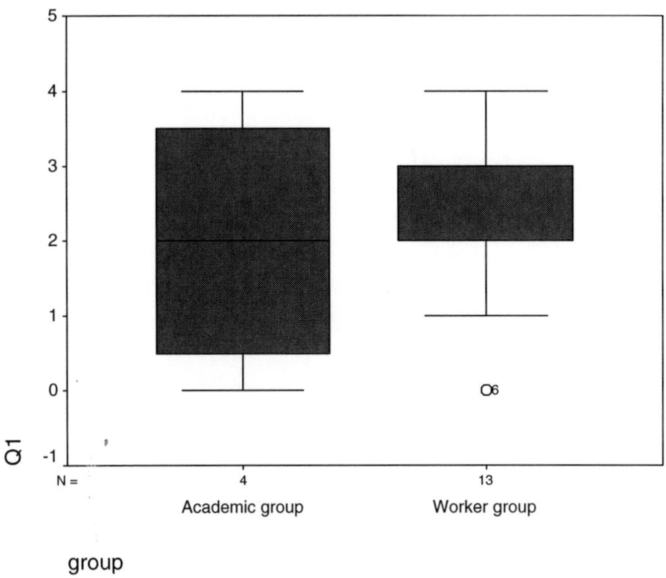

Q2

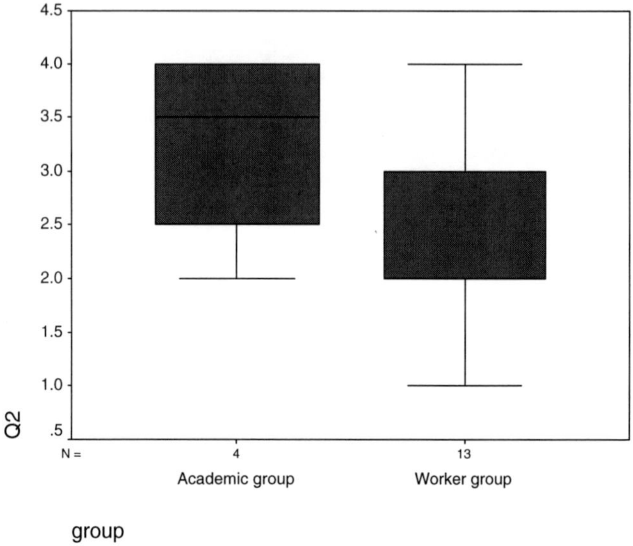
4.5
4.0
3.5
3.0
2.5
2.0
1.5
1.0
.5
Q2
N =
4
13
Academic group
Worker group
group

Q3

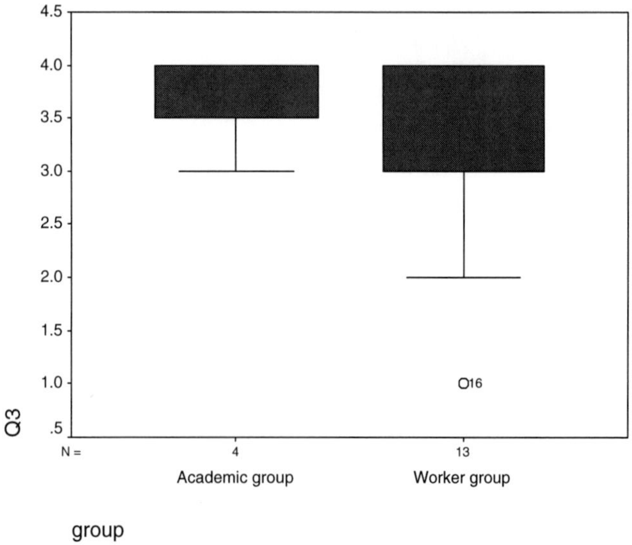
4.5
4.0
3.5
3.0
2.5
2.0
1.5
1.0
.5
Q3
O16
N =
4
13
Academic group
Worker group
group

Q4

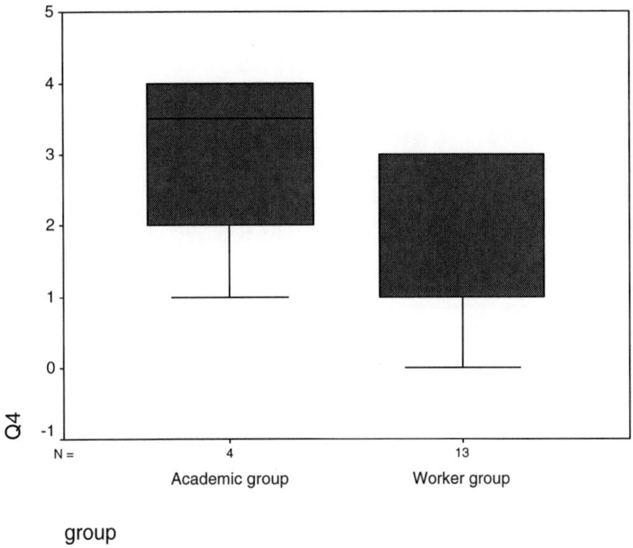
5
4
3
2
1
0
-1
Q4
N =
4
13
Academic group
Worker group
group

Q5

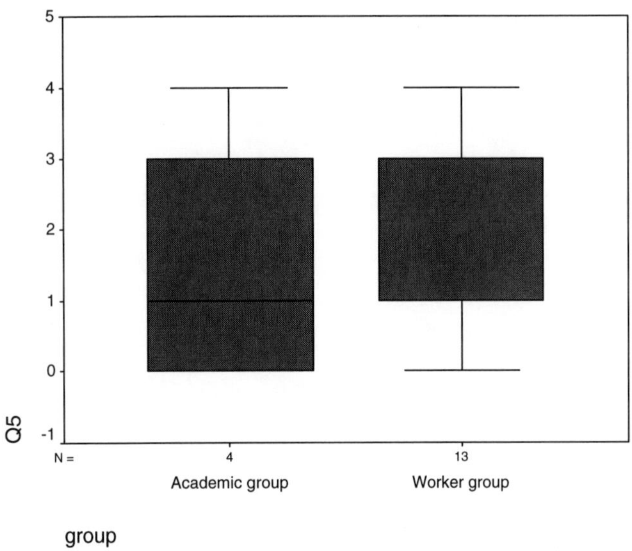
5
4
3
2
1
0
-1
Q5
N =
4
13
Academic group
Worker group
group

Q6

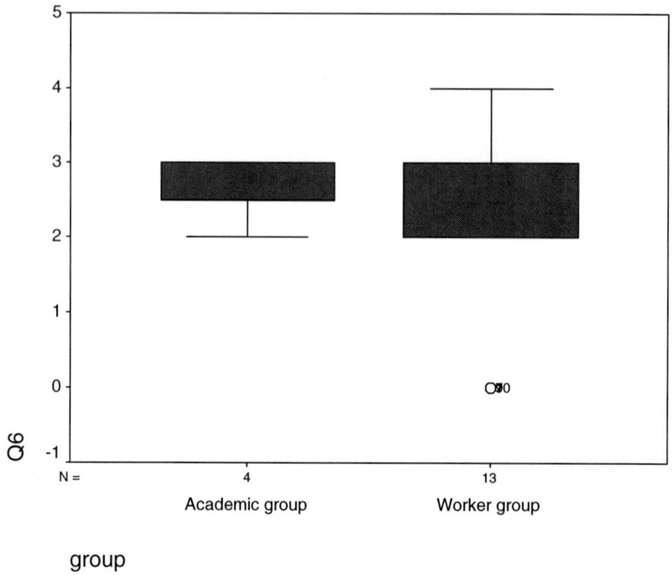

In the regression for the effect of management commitment on marketing results and CSI, it was found that there was a significant effect (p= 0.039), by effect of 56.4%.

In the regression for effect of ethics on marketing results and CSI, it was found that there was no significant effect (p= 0.601).

6-1-3 Action-Results Outcome of the focus group

The biggest effects of the Action-result matrix were as follow. The numbers in the boxes represent the numbers of votes:

Table 6-3

	Cost Cuts	Corp. values	Team Performance	COQ	Marketing results	CSI
Top management commitment		7 √	√ 8			
Attitude & Behaviour			√ *12*			
Safety	√ 8					
Impact on society		√ 8				
Employee motivation			√ 11			
Training and development			√ 9			
Improving Productivity	**√ 7**			**√ 7**		
Marketing results and competitiveness					**√ 9**	**√ 11**

The biggest result hoped to be achieved in Egypt from the initiators (Actions) is the team performance, therefore it is upgraded in the HEQMA scoring to reflect the focus group tendency.

The highest action-result ratio is the "Attitude and behaviours – Team performance". It is natural correlation on the matrix.

6-2 Holistic Egyptian Quality Management Approach, HEQMA

1- Base factor:

1-1 Top management commitment 14 %

1-2 Corporate Ethics 13 %

2- Human capital factor:

2-1 Training and development 12 %

2-2 Team performance 11 %

2-3 Employee motivation 6 %

3- Internal Capability Factor:
3-1 Improving Productivity 7 %
3-2 Cost of Quality 7 %
3-3 Safety and positive impact on society 8 %
4- Outcome factor:
4-1 Marketing results and competitiveness 11 %
4-2 Customer Satisfaction Index, CSI 11 %

Table 6-4: Holistic Egyptian Quality Management Approach, HEQMA

TQM parameters and Mechanisms	% /no
1- BASE FACTOR: **1 – 1: Top Management commitment:** 1 - Defining a customized and unique mission and vision, and ensuring commitment by all staff. 2 - Describing and justifying strategy and policies. 3 - Criteria for decision process: Fact-Based, improvement or business fire fighting, Validity of decisions taken recently. 4 - Types of board decisions; Impact on technology transfer. 5 - Interference of Management Information Systems in decisions. 6 - Effectiveness of decisions taken on providing/improving resources. 7 - Percentage of use of external consultants (not affiliated to the organization) in any business process development. 8 - Adopting networking as a valuable source of knowledge. 9 - Acceptance and implementation of self-assessment and this model half yearly. 10 - Percentage of accepted suggestions related to proposed ones. 11 - Succession plans for 20% of board members, percentage leaders developed. 12 - Annual Rotation plans for line managers.	
1 – 2: Corporate Ethics: Values, Attitudes and Behaviours: 1- Management can describe/ensure the equality of opportunity in the selection and development of people. 2 - Relevance of decisions' and activities' values to business mission. 3 - Managers Pace making practices, verified by anonymous persons. 4 - Defined criteria for recruitment based on values and efficiency. 5 - Conducting top-middle-supervisory meetings efficiently; Weekly. 6 - Two-way communications on the "minutes of meetings" records. 7 - Conducting out-of-office (field) events Quarterly. 8 - Percentage of theft/unjustified absence cases. 9 - Awareness of soft skills and conflict management for all line managers and supervisors. 10 - People can explain the aims and objectives of their department.	

2 – HUMAN CAPITAL FACTOR **2 - 1: Training and development:** 1 - Percentage of training hours, related to number of employees, spent on: OJT – Apprenticeships – Internal demos. 2 - Measuring the return on training: the following action plan - reports on training effects on business results; increase in employee response and cooperation – less employee complaints – percentage of tasks done on time – improved decision making process – higher innovation rates. 3 - Updated competence measures, semi-annual. 4 - Percentage rise in: Knowledge, Skills and Attitude (KSA) for each employee. 5 - Evidence of top management support for people's development; including conducting training needs assessment (TNA). 6 - Managers can describe examples to encourage people to improve.	
2 - 2: Team performance: 1 - Measuring the effectiveness of processes due to teams' recommendations. 2 - Availability of agreed standard contracts for customers and suppliers. 3 - Operating forms compliance and integration between departments. 4 - Moral and/or financial recognition for teams based on results; Teams can describe positively how their contribution was recognized. 5 - Updated database, companywide, semi-annual. 6 - Percentage of benefits to employees from applying management system. **2 - 3: Employee motivation:** 1 - Competence-income relation, based on individual results. 2 - 360 degree feedback for departmental performance. 3 - Availability of job security plans, decreased manpower turnover ratio. 4 - New employees confirm technically regarding the effective induction. 5 - Managers can describe the costs and benefits of the development of people. 6 - Employee income is amongst the top 20% in the same field. 7 - Measuring of the authority-responsibility equity.	
3 – INTERNAL CAPABILITY FACTOR **3 – 1: Improving Productivity:** 1 - Availability of time studies for processes and improving their percentage. 2 - Percentage of productivity increase by modifications made to production lines. Managers can show the improved processes. 3 - Percentage of inserting technology transfer into production. 4 - Number of R&D employees in production facility. 5 - Revenues of improved products as a percentage of total revenues. 6 - Percentage of new products/projects using available resources. 7 - Life cycle of new products/improvements. 8 - Percentage conformity of suppliers' offers. 9 - Managers can describe the cost and benefits of the development of internal processes.	

10 - Measuring and increasing man-hr productivity. 11 - Using product sampling plan, where appropriate. **3-1-1: Cost cuts** 1 - Percentage of Lean practices in: product development, procurement, manufacturing, distribution, and the lean enterprise. 2 - Availability of computerized Central Inventory Control. 3 - Efficient benchmarking in materials/manpower/time, as appropriate. 4 - Managers can give enough examples of relevant and timely cuts. **3-1-2: Waste reduction:** 1 - Classifying number of wastes: Time/machines/materials/manpower 2 - Reducing waste percentage of total inputs, including non-conformities of suppliers. 3 - Raising acceptance levels of suppliers' materials. 4 - Availability of analysis system for cost of materials. 5 - Improving depreciation levels of machines. **3 – 2: Cost of Quality:** 1 - Improving the percentage of preventive cost related to failure and appraisal; Reporting quarterly. 2 - Managers can describe specific actions to support minimizing cost. 3 - Managers can describe the costs and benefits of the development of costs of quality. **3 – 3: Safety and Impact on Society:** 1 - Awareness of safety measures; 2 - Percentage increase in adherence to standard safety measures, audited by independent appointees. 3 - Percentage decrease in yearly accidents or incidents. (Safety includes health measures and dress code). Reported quarterly. 4 - Percentage of projects to surroundings of total sales, aside from harming surroundings. Positive and friendly environmental effects and impacts. 5 - Percentage of total sales to awareness and quality of life training to surrounding society. 6 - Percentage of employees of surrounding society, distributed to all functions. 7 - Percentage of recycled materials in production.	
4 – OUTCOME FACTOR **4 – 1: Marketing results and competitiveness:** **4-1-1: Competitiveness and customer focus:** 1 - Percentage of price difference related to main competitors by benchmarking to the same product class. 2 - Percentage of reaching and covering competitors' territories. 3 - Offering different payment options and/or diversified dealings for all pricing items. 4 - Relation between marketing costs and/or campaigns and revenues. 5 - Company share of customer purchases and/or company's market share. 6 - Percentage of customers transferred from competitors to company.	

7 - Concerned managers can explain what they do to support quality marketing. **4-1-2: New market segments:** 1 - Percentage of new annual coverage. 2 - Diversification of products to fulfil new areas and/or customers' buying habits. 3 - Investment in R&D related to exploring new customers/markets/products. 4 - Training hours of above employees to seek opportunities of new business. **4-1-3: Export increase target:** Same parameters as "new market segments", projected to int'l markets.	
4 – 2: Satisfied Customers Index: 1 - Percentage of sales increase to top 10 customers. 2 - Repeatability of sales orders on different products. 3 - Percentage increase of new to old customers. 4 - Annual decrease in number of customer complaints. 5 - Records of percentage increase of long term deals with (classes of) customers.	

6-3 Conclusions

The main contributions that this research has achieved are considered below:

- Identifying HOW TO MEASURE different quality elements in detailed attributes, e.g. how to measure Q.C. to reach and sustain consistency of product quality; how to succeed through people by measuring the team performance, training and development, and motivation.
- Defining the main quality elements for EIO in four bundles; while integrating them in a unique way.
- Calling for other business attributes in quality concepts, such as financial, safety, and marketing perceptions.

There were special traits in this HEQMA for the Egyptian culture. Top management and corporate ethics have special characteristics, as explained in the focus group and obstacles to HEQMA (chapter 4), for HR capital, group work and teamwork as a substitute for charismatic individual leaderships. The relation between mission, vision, values and strategy has a separate appendix.

For the Egyptian industry, the special trait is the sequence of COQ and Q.C. which needs great effort before introducing process management, the current general concept.

It is suggested that, even though quality is considered to be an important issue, it is not fully appreciated by organizations, and only some of them use the quality management system strongly. Companies usually do have quality systems, but approach MPI in many other ways, such as finance and marketing.

Nevertheless, using HEQMA as a means to drive MPI, EIO can achieve the required results with integrative techniques.

The percentage of contribution in HEQMA indicators are as follows:

43% from the researcher's background information and fieldwork through the years, 33% from different QM models, 17% from ideas from references, and 7% are from focus group comments. The numbers only indicate quantitative measures; otherwise the researcher puts the most value on the parental participation from the research study supervisor, and the positive additions from the focus group interviewees. It is a strong effect of the focus group to correct the weights of the items, and most importantly to readjust the indicators to reflect the Egyptian culture.

Analysis and interrelation never ends. It is enough here to know the general trends corresponding to the content analysis.

It is expected that HEQMA will prove to be a success in the Egyptian industrial climate, as it fulfilled the methodology for building up a model. It also provides methods of quality implementations, not only principles. HEQMA can be helpful in corporate governance, sustainable development and transparency.

6-4 Limitations

The normal limit of the study population is the Egyptian manufacturing organization within Egypt. Time limits are still valid till further implementation or modifications of the model.

1 - There is a need to highlight that the work in this study is not intended to produce a QMS standard, but rather to produce an APPROACH derived from TQM entries and principles, from the essence of the models that already exist and from EIO culture.

Leonard and McAdam (2002) also explained the same.

2 - There is a concern, an interviewee urged, about a possible defacement of the Egyptian personality, implied in the focus group, which is reflected in the work values, such as quality, ambition, honesty, order, strive, success and time factor. These would of course affect the model

implementation; results depend on the truth and attitudes of the implementers.

3 - It is beyond the scope of the current research to deeply investigate the implementation of HEQMA; this is suggested as a possibility for further research, which will depend on each sector's needs as well.

It is recognized that further research is necessary to establish the exact nature of the causal linkages between performance measures and strategic intent, and also to gain insights into practice elsewhere, as Evans (2005) expressed.

Zairi et al (1994) assumed that it would take some years for TQM implementation to start to yield positive results.

4 - This approach will be helpful also when the organization proceeds for a specified quality management standard. There are three types of organizations in this respect: one which seriously adopts a certain QMS would be distracted when experiencing this HEQMA; in that case it is better for them to continue without interfering with HEQMA. There are others that still do not appreciate QMS values; they are the certified companies that were originally seeking certification for awards only. Therefore, it is encouraging in this instance to apply HEQMA as a whole. On the other hand, there are many leading and highly profitable Egyptian organizations who implement independent managerial principles and their own measures of performance rather than adopting a certain QMS; they can then closely regard the items of HEQMA, which can benefit them.

5 - EIO does not need to be criticized or diagnosed in the formulation of HEQMA, as they change from one business sector to another, and from time to time through the years.

6 - Finally, HEQMA is not supposed to cover the service industry, which has many different characteristics than manufacturing, such as responsiveness-customer contacts including ease-courtesy, etc. (Ghobadian et al 1993).

7 - Risk management and SWOT analysis were not conducted on this model, as Egyptian industries may not yet be able to reach those horizons. A further deep independent study could be of interest for other researchers.

6-5 Recommendations

The model of this study is derived from versatile international and accredited quality concepts, but matching Egyptian climate and business culture.

1 - The privileged signs of the model (e.g. measurable sub-items), as well as its low cost compared to other models, its importance and its criticism make the difference, as the essence of the PhD is Modelling;
2 - If the proposed model could not solve the many problems EIO are facing currently, it is hoped that it can help solve some of them;
3 - At least it can clarify the road where problems are laid;
4 - It can also draw the attention of the concerned parties to the proper channel for diagnosing problems;
5 - Many papers which were reviewed recommend that new researchers should modify or develop new TQM models and/or approaches;
6 - As a principal need for the model, the availability of many national models in other countries provided the necessity of composing a national model for EIO.

Based on the literature review and focus group interviews, it is recommended:

1 - To investigate the possibility of applying the Japanese concept of employees' partial possessions of assets to enforce self-motivation and MPI.
2 - The expected added value of HEQMA is when it is robust; i.e. there is a need to minimize the "Effect of External Variables; PEST" which is usually difficult to control.
3 - Other than questions and answers revealed in the focus group, HEQMA must be part of continuous planning of business. It is evolved essentially by consultation, then by using the expertise inside each organization.
4 - The organization applying HEQMA can slightly determine/change the weight for each of the HEQMA elements. The criteria depend on the specific business needs, the TQM status, and the maturity of the company/business.
5 - It is essential to include the role of HR in the implementation of HEQMA, i.e. to provide competent HR who can cover all items, rather than a Quality department in the normal QMS. It has been argued that TQM items included in HEQMA cannot be applied in isolation, as TQM is a total philosophy involving all organization members and has high HR content.
6 - The paper will be useful to academics with an interest in implementation of performance measurement, and also to practitioners

seeking an understanding of other practical managerial tools. However, HEQMA is subject to intellectual property and patent rights.

7 - It is important to create domestic high technology, and important to modernize the Egyptian industries, which leads to a competitive advantage.

APPENDICES

Appendix A

Detailed statistical Analysis

Regression for effect of Management commitment on Marketing results, and CSI

Variables Entered/Removed(b)

Model	Variables Entered	Variables Removed	Method
1	Mktg results, CSI(a)	.	Enter

a All requested variables entered.
b Dependent Variable: Mgt comm

Model Summary

Model	R	R Square	Adjusted R Square	Std. Error of the Estimate
1	.564(a)	.318	.220	3.99685

a Predictors: (Constant), Mktg results, CSI

ANOVA(b)

Model		Sum of Squares	df	Mean Square	F	Sig.
	Regression	104.118	2	52.059	3.259	.069(a)
	Residual	223.647	14	15.975		
	Total	327.765	16			

a Predictors: (Constant), Mktg results, CSI
b Dependent Variable: Mgt comm

Coefficients(a)

Model		Unstandardized Coefficients		Standardized Coefficients		
		B	Std. Error	Beta	t	Sig.
1	(Constant)	6.521	3.425		1.904	.078
	CSI	.118	.265	.103	.445	.663
	Mktg results	.583	.256	.525	2.280	.039

a Dependent Variable: Mgt comm

From this model it was found that there was a significant effect of Management commitment on CSI (p 0.039), by effect of 56.4%

Regression for effect of Corp. Ethics on Marketing results, and CSI

Variables Entered/Removed(b)

Model	Variables Entered	Variables Removed	Method
1	Mktg results, CSI(a)	.	Enter

a All requested variables entered.
b Dependent Variable: Ethics

Model Summary

Model	R	R Square	Adjusted R Square	Std. Error of the Estimate
1	.309(a)	.095	-.034	7.48483

a Predictors: (Constant), Mktg results, CSI

ANOVA(b)

Model		Sum of Squares	df	Mean Square	F	Sig.
1	Regression	82.742	2	41.371	.738	.496(a)
	Residual	784.317	14	56.023		
	Total	867.059	16			

a Predictors: (Constant), Mktg results, CSI
b Dependent Variable: Ethics

Coefficients(a)

Model		Unstandardized Coefficients		Standardized Coefficients	t	Sig.
		B	Std. Error	Beta		
1	(Constant)	19.629	6.414		3.060	.008
	CSI	-.442	.496	-.237	-.891	.388
	Mktg results	-.256	.479	-.142	-.535	.601

a Dependent Variable: Ethics

From this model it was found that there was no significant effect of Corp. Ethics on CSI or Marketing results.

Reliability

****** Method 2 (covariance matrix) will be used for this analysis ******

R E L I A B I L I T Y   A N A L Y S I S - S C A L E (A L P H A)

Covariance Matrix

	VAR00002	VAR00003	VAR00004	VAR00005	VAR00006
VAR00002	15.4926				
VAR00003	4.6140	16.6397			
VAR00004	-3.5515	-7.5147	10.1397		
VAR00005	-1.5699	-3.7610	2.1985	9.4926	
VAR00006	2.2574	-6.9265	3.3640	2.2574	18.9926
VAR00007	-4.4779	-.7169	3.7794	4.5221	7.1654
VAR00008	-5.6544	-7.7316	2.7316	2.5331	-2.0331
VAR00009	-1.9154	1.5956	-3.8456	-1.9154	-7.0846
VAR00010	-8.0368	-6.3051	.7426	-12.4118	-13.1507
VAR00011	4.5184	10.2463	-5.8088	-1.0441	-4.3934

	VAR00007	VAR00008	VAR00009	VAR00010	VAR00011
VAR00007	11.3088				
VAR00008	-.2243	9.8824			
VAR00009	-5.9412	3.3382	9.9338		
VAR00010	-15.3272	3.8529	9.7978	54.1912	
VAR00011	-.1176	-7.6140	-4.6176	-10.9706	20.4853

N of Cases = 17.0

Analysis of Variance

Source of Variation	Sum of Sq.	DF	Mean Square	F	Prob.
Between People	10.0235	16	.6265		
Within People	3513.6000	153	22.9647		
Between Measures	698.6824	9	77.6314	3.9713	.0002
Residual	2814.9176	144	19.5480		
Nonadditivity	1.0486	1	1.0486	.0533	.8178
Balance	2813.8690	143	19.6774		
Total	3523.6235	169	20.8498		
Grand Mean	10.0471				

Tukey estimate of power to which observations must be raised to achieve additivity = -.6030

R E L I A B I L I T Y A N A L Y S I S - S C A L E (A L P H A)

Reliability Coefficients 10 items

Cronbach Alpha =-0.756 Standardized item alpha = -7.0009

Reliability

****** Method 2 (covariance matrix) will be used for this analysis ******

R E L I A B I L I T Y A N A L Y S I S - S C A L E (A L P H A)

Covariance Matrix

	VAR00012	VAR00013	VAR00014	VAR00015	VAR00016
VAR00012	1.8676				
VAR00013	.2169	1.2574			
VAR00014	.2243	-.0221	.8162		
VAR00015	.1875	.0000	.6250	1.5000	
VAR00016	.3934	-.1985	.3456	.0000	1.9853
VAR00017	-.2353	.1287	.0515	-.1875	.3382

	VAR00017
VAR00017	1.5956

N of Cases = 17.0

Analysis of Variance

Source of Variation	Sum of Sq.	DF	Mean Square	F	Prob.
Between People	34.0196	16	2.1262		
Within People	127.0000	85	1.4941		
Between Measures	16.6667	5	3.3333	2.4169	.0430
Residual	110.3333	80	1.3792		
Nonadditivity	.2029	1	.2029	.1456	.7038
Balance	110.1304	79	1.3941		
Total	161.0196	101	1.5943		
Grand Mean	2.4314				

Tukey estimate of power to which observations must be raised to achieve additivity = 1.4646

Reliability Coefficients 6 items

Cronbach Alpha = .7514 Standardized item alpha = .3812

The alpha in the two model of reliability was more than 0.7. i.e. there was stability of the questionnaire and was reliable

CSI

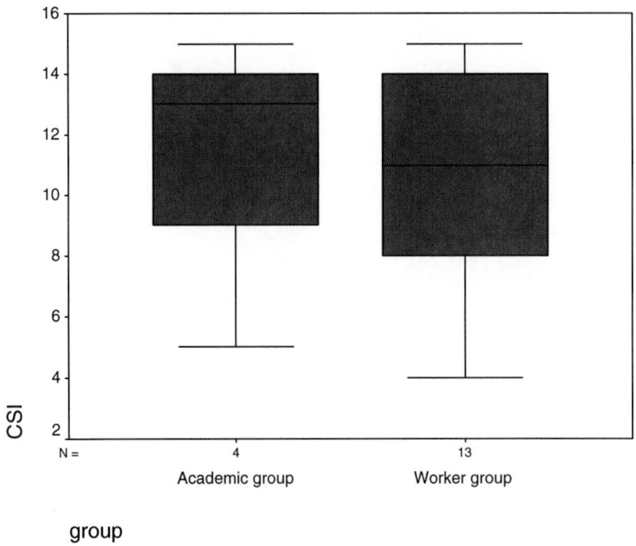

Marketing results

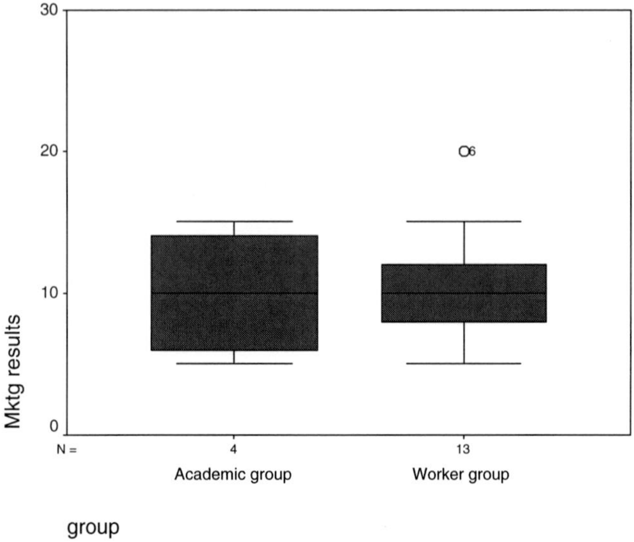

Safety and Society effects

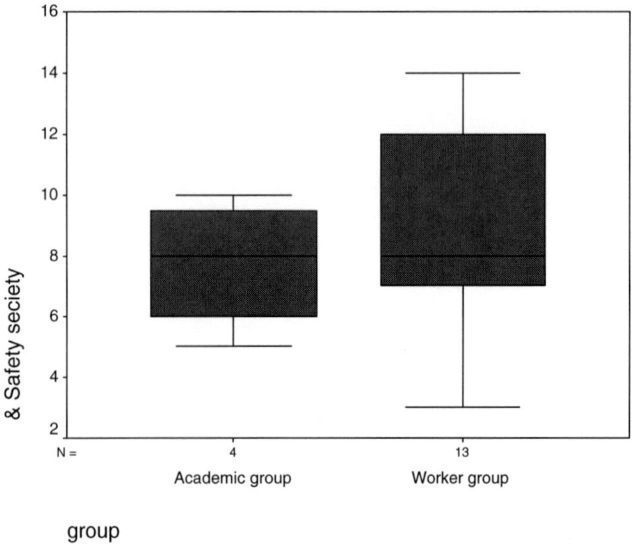

CoQ

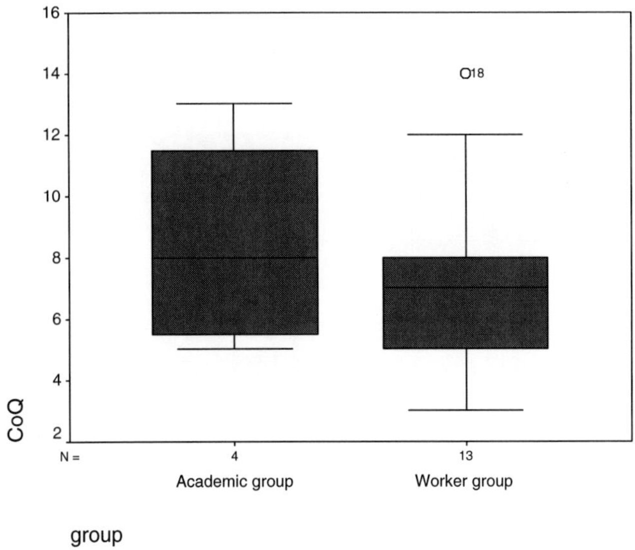

Improving productivity

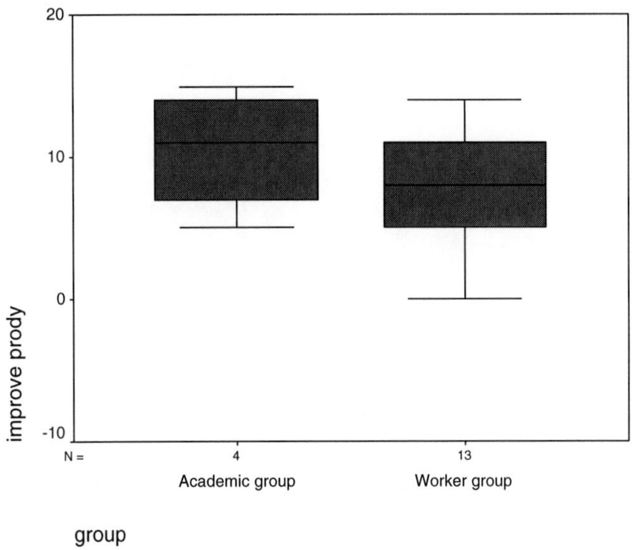

Motivation

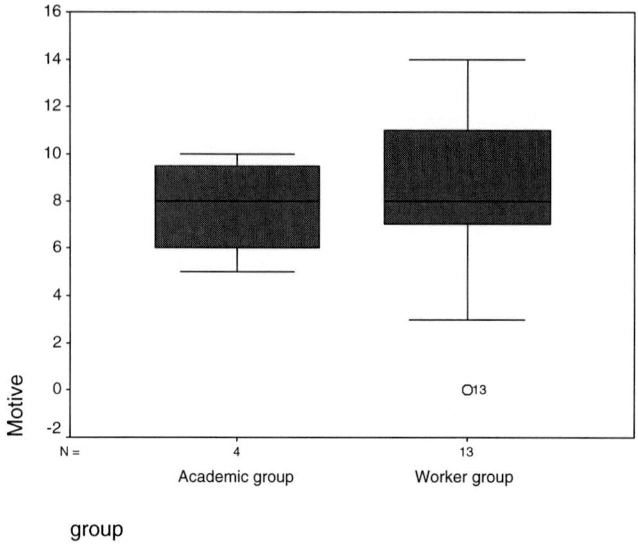

Team performance

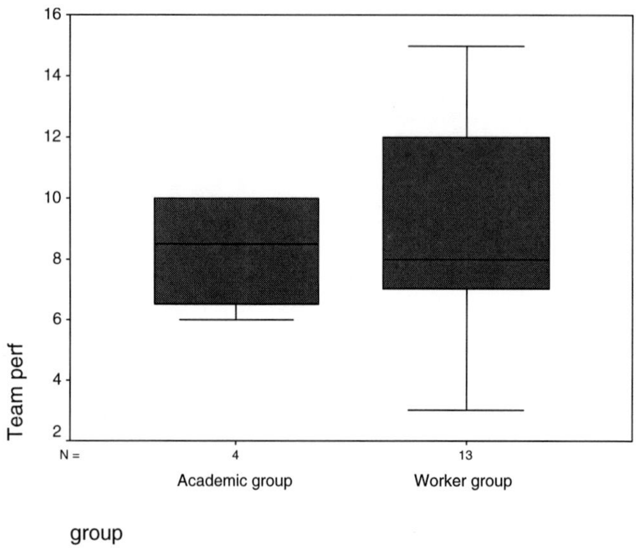

Training and development

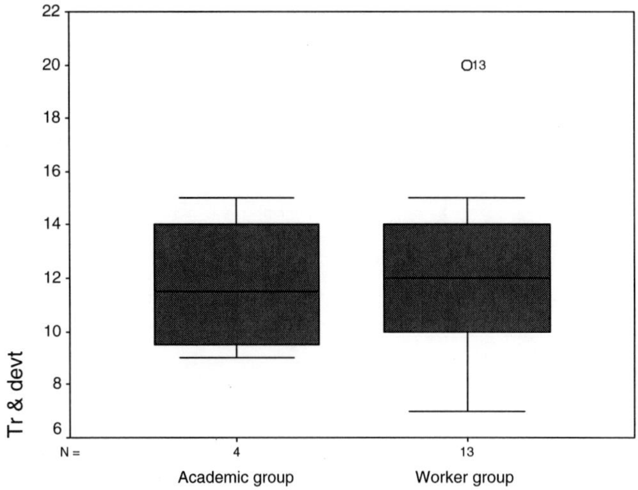

group

Ethics

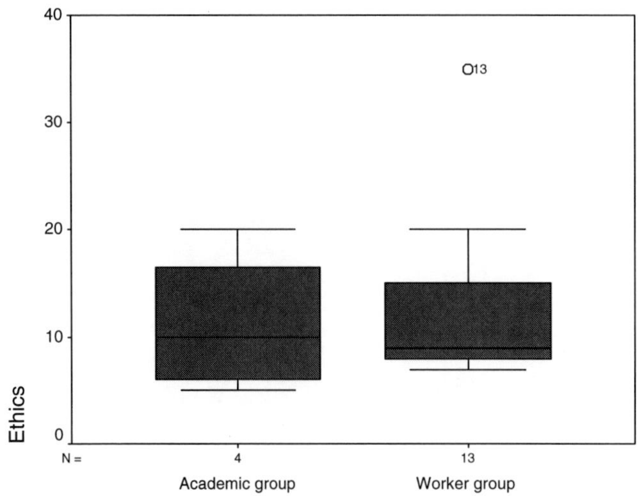

group

Appendix B

Characteristic features of the studied sample.

	Pirelle	Pharco	America Pharma-cueticals	Al-Arabi lighting technology	Al-Arabi home appliances	Egyptian Petro-chemicals
Working years	<5	21-35	21-35	<5	> 40	21-35
No. of Employees	500-1000	>1000	>1000	500-1000	100-250	>1000
Info sys advance	Database &knolw	Database &knolw	Traditional manual data	Database &knolw	Database &knolw	Separate files
Marketing strategy	Differentiation	Focus strategies	Cost leadership	Differentiation	Differentiation	Focus strategies
Performance measures	- Financial measures - Customer satisfaction - Market share	- Market share	- Financial measures	- Financial measures - Customer satisfaction - Market share - Defect percentage	- Financial measures - Customer satisfaction - Market share - Defect percentage	- Financial measures - Customer satisfaction - Market share - Defect percentage

BIBLIOGRAPHY

1. Crosby, Philip B (1979) "quality is free", McGraw-Hill, USA
2. Philips and Stone, 2006, www.edara.com, August 2006, How to measure Training results Mc Graw Hill (ISBN: 007138792-7)
3. Juran, J.M. and Gryna, Frank (1993), quality planning and analysis, McGraw-Hill international edition.
4. Kaplan and Norton, 2002, Strategy Maps: Converting intangible assets into tangible outcomes Harvard Business School Press 2004, Pages: 450
5. Leeman, Choy, Jan 2002: Development and validation of a TQM for services, UMI Microform Edition, 3038000 Pro-request Information and learning company.
6. Mussalaam, Ali, Strategic HR management, 2007, Alexandria University.
7. Pekgun et al, 2006, Coordination of marketing and production for price and lead time decisions IIE Transactions (2008) 40, 12–30 ISSN: 0740-817X print / 1545-8830 online DOI: 10.1080/07408170701245346
8. Armitage, 2002, Business Excellence model in SME. Managerial auditing Journal, 17-1-2 [2002] 26-35
9. Agarwal and Teas, 2002, Marketing Quality Awards. Journal of product and brand management, Vol. 11 No.4 2002
10. Al-Mashari et al, 2001, BPR, International Experience. Business Process Management Journal, Vol.7 No.5 2001 PP 437-455
11. Aoki, 2007, transferring Japanese kaizen activities to overseas plants in china; International Journal of Operations &Production Management Vol. 28 No. 6, 2008 pp. 518-539
12. Assiri et al, How to profit from BSC, implementation roadmap Industrial Management & Data Systems Journal, Vol. 106 No. 7, 2006 pp. 937-952
13. Boys et al, 2003, ISO 9004: a path to business excellence International Journal of Quality & Reliability Management Vol. 21 No. 8, pp. 841-860
14. Doolen et al, 2008, kaizen events and organizational performance: a field study International Journal of Productivity and Performance Management Vol. 57 No. 8, 2008 pp. 637-658

15. Elsey and Fujiwara, 2000, Kaizen and technology transfer instructors as work-based learning facilitators in overseas transplants. Journal of Workplace Learning Volume 12. Number 8. 2000. pp. 333±341
16. Eraqi, Muhammad, Tourism services quality in Egypt, Benchmarking: An International Journal Vol. 13 No. 4, 2006 pp. 469-492
17. Ghobadian et al 1993, Service quality, concepts and models International Journal of Quality & Reliability Management, Vol.11 No.9,1994 pp. 43-66,
18. Ismail, 2007, BSC Performance evaluation measures in the private sector: Egyptian practice Managerial Auditing Journal, Vol. 22 No. 5, pp. 503-
19. Karlsson and Ahlström 1995, change processes towards lean production: the role of the remuneration system International Journal of Operations & Production Management, Vol. 15 No. 11, 1995, pp. 80-99.
20. Katayama and Benett, lean production in a changing competitive world: a Japanese perspective International Journal of Operations & Production Management, Vol. 16 No. 2 1996, pp. 8-23
21. Lee and Peccei, 2006 lean production and quality commitment, Personnel Review Journal, Vol. 37 No. 1, 2008 pp. 5-25
22. Lluster et al, 2005, To what extent do enablers explain results in the EFQM excellence model? An empirical study International Journal of Quality & Reliability Management Vol. 22 No. 4, 2005 pp. 337-353
23. Magd, 2008, Understanding benchmarking in Egyptian organizations: Bench Marking: An International Journal Vol. 15 No. 6, 2008 pp. 742-764
24. Magd, 2006, ISO 9001:2000 in the Egyptian manufacturing sector: perceptions and perspectives, e-TQM College, Dubai International Journal of Quality & Reliability Management Vol. 25 No. 2, 2008 pp. 173-200
25. McAdam 1999, quality models in an SME context, A critical perspective using a grounded approach. International Journal of Quality & Reliability Management, Vol. 17 No. 3, 2000, pp. 305-323.
26. Mostafa, 2005, An empirical study of patients' expectations and satisfactions in Egyptian hospitals, International Journal of Health Care Quality Assurance Vol. 18 No. 7, 2005 pp. 516-532
27. O'Neill and Sohal, International Journal of Operations & Production Management, Vol. 18 No. 9/10, 1998, pp. 832-864,
28. Olivella et al, 2007, work organization practices for LP Journal of Manufacturing Technology Management Vol. 19 No. 7, 2008 pp. 798-811

29. Park, 2008, Excellence models and management control The TQM Journal Vol. 20 No. 2, 2008 pp. 98-119
30. Raisinghani 2005, six sigma: concepts, tools, and applications Industrial Management & Data Systems Journal, Vol. 105 No. 4, 2005 pp. 491-505
31. Salaheldin and Eid, 2007, the implementation of WCM in Egyptian manufacturing firms; Industrial Management & Data Systems Journal, Vol. 107 No. 4, 2007 pp. 551-566
32. Salaheldin, 2005, JIT implementation in Egyptian manufacturing firms: some empirical evidence International Journal of Operations & Production Management Vol. 25 No. 4, 2005 pp. 354-370
33. Schiffauerova and Thomson, 2005, A review of research on cost of quality models and best practices International Journal of Quality & Reliability Management Vol. 23 No. 6, 2006 pp. 647-669
34. Sinclair, David and Zairi, Mohamed (1995) "Effective process management through performance measurement: Part III-an integrated model of total quality-based performance measurement", Business Process Management Journal, Vol. 1 Iss: 3, pp.50 – 65
35. Stebbins et al, Journal of Organizational Change Management, Vol. 11 No. 3, 1998, pp. 216-232,
36. Wiele et al, 2000, ISO 9000 series registration to business excellence: the migratory path Business Process Management Journal, Vol. 6 No. 5, 2000, pp. 417-427
37. Zein Eldin, 2006, The quality of health care and patient satisfaction International Journal of Health Care Quality Assurance Vol. 19 No. 1, 2006 pp. 60-92
38. Almukaadem, Mustafa (2009) interactive TQM effects, Egyptian services, PhD Thesis.
39. ASQ magazine, 3/2003. Page 45 & 89
40. B.A.G. Bossink et al, 1992, Diagnosing TQM, VOL. 3, NO. 3, 1992
41. British Standards Institute BSI, 1992, TQM guidelines, part 1,
42. British Standards Institute BSI, 1992, TQM guidelines, part 2
43. Conti, Teto 1999, ISO, Vision 2000: positioning the new ISO 9000 standards with respect to total quality management models; TOTAL QUALITY MANAGEMENT, VOL. 10, NOS 4&5, 1999, S454± S464
44. Abdel Maksoud, El-Sayed, Cost & Benefit Analysis, PHD course, 2007
45. Elaraby and Abdelghany, 2008, TQM: from where to where, AASTMT
46. El-Kott, Ghada, Research methodology, Lectures notes (2008)

47. Freisleben, 2006, communicating 6 sigma to top management. MEASURING business excellence VOL. 10 NO. 2 2006 PAGE19- 27
48. Gautreau and Kliener, 2001, BSC recent trends in performance Evaluation Management Research News, Volume 24 Number 3/4 2001
49. Hides et al, 2004, EFQM self assessment The TQM Magazine, Volume 16 No. 3, 2004, pp. 194-201
50. ISO 9001: 2008 Quality management system standard
51. Louise, Lectures on Advanced Studies in Quality Management, 1999
52. Labib 2006, Quality Management in higher education institutions, AASTMT.
53. Magd and Curry, 2003, An empirical analysis of management attitudes towards ISO 9001:2000 in Egypt The TQM Magazine Volume 15, Number 6, 2003 pp. 381-390
54. Sallah Eldin, 2003, The implementation of TQM strategy in Egypt, a field-force analysis, The TQM magazine, Vol. 15 No. 4 2003, pp. 266-274
55. Tennant and Wu, The TQM Magazine Vol. 17 No. 6, 2005 pp. 537-545
56. Vouzas, Fotis, 2007, Investigating the human resources context and content on TQM, business excellence and ISO 9001:2000 MEASURING BUSINESS EXCELLENCE VOL. 11 NO. 3 2007, pp. 21-29,
57. Zairi, M, et al, (1994) "Does TQM Impact on Bottom-line Results?" The TQM Magazine, Vol. 6 Iss: 1, pp.38 - 43
58. coer.org.nz National Quality / Business Excellence Awards in different countries (Date of research: 9-13 January 2010: Centre for Organizational Excellence Research)
59. Dti.gov.uk/quality/people, dept. of trade and industry; For business innovation and skills (focus group questions).
60. Dubaided.gov.ae
61. eos.org.eg, 1/2010. Code of Ethics for standards and compliance assessment, Egyptian organization for standardization and quality. (pp1-39)
62. EFQM.org 3/2010
63. Geert-hofstede.com 2/2010
64. iso.org, ISO, 2001,10/2009
65. imc-egypt.org 4/2010
66. Kaaps.jo 4/2010 Jordanian trade and industrial ministry
67. nqiegypt.org, 3/2009
68. nationalawards-eg.com 4/2010
69. thequalityportal.com 5/2010.

70. 2/2010 70. Saso.org.sa
71. wtec.org 5/2010 world technology evaluation centre

Arab Republic of Egypt Ministry of Scientific Research Academy of scientific Research & Technology PATENT OFFICE /WIPO	EGYPTIAN PATENT OFFICE مكتب براءات الاختراع المصري	جمهورية مصر العربية وزارة الدولة لشئون البحث العلمي أكاديمية البحث العلمي والتكنولوجيا مكتب براءات الإختراع

محضر ايداع مستندات طلب براءة اختراع

التاريخ : ٢٠١٢/١١/١٩ الوقت :	رقم الإيصال :	رقم الطلب : ٢٠١٢/١٩١٨
الجنسية : مصري	إسم الطالب : اسامه أحمد كمال عبد الغني	

مسمى الاختراع : منظومة المعايير المصرية الشاملة للجودة Holistic Egyptian Quality Management Approach

أسم الوكيل (إن وجد): نقطة اتصال جامعة الاسكندرية	بيان المستندات المستلمة مع الطلب

- [x] 1 - نموذج طلب براءة اختراع أو شهادة منفعة HEQMA
- [x] 2 - وصف تفصيلي للاختراع
- [x] 3 - الرسم الخاص بالاختراع إذا كان ضروريا إدراك الاختراع أو كان طابع الاختراع يسمح بذلك
- [x] 4 - ملخص الاختراع باللغة العربية والانجليزية مصحوبا بأفضل رسم توضيحي إن وجد
- [ ] 5 - مستخرج من السجل التجاري أو مستخرج رسمي من عقد التأسيس إذا كان الطالب شركة أو هيئة
- [x] 6 - سند الوكالة إذ أودع الطلب بواسطة وكيل
- [ ] 7 - المستند الدال على أحقية الطالب في الاختراع إذا كان الطالب غير المخترع (التنازل)
- [ ] 8 - موافقة صاحب الشأن إذ كانت العناصر الجوهرية للاختراع قد تم الحصول عليها من اختراع شخص آخر
- [ ] 9 - إذا كان الطلب يتضمن الرغبة في اعتبار الأولوية في التسجيل لطلب سبق تقديمه في دولة تكون طرفا في اتفاقية أو معاهدة دولية مع دولة جمهورية مصر العربية وفقا للمادة (11) من القانون فانه يجب تقديم صورة من الطلب السابق و المستندات المرفقة به مصحوبة بشهادة تبين تاريخ و رقم إيداعه و الدولة التي أودع فيها.
- [ ] 10 - مستندات طلب (PCT) المنشور و تقرير البحث و الفحص الفني
- [ ] 11 - الشهادة الصادرة بالحماية المؤقتة إن وجدت
- [x] ١٢ - تعهد كتابي بتقديم اللازم من المستندات عدا المرفق بالطلب منها (من 4 - 11)
- [x] 13 - اخرى

مجموع المستندات المستلمة: ١١	اسم المستلم :
الختم :	التوقيع : ٢٠١٢/١١/٢٩

يؤشر بعلامة ☒ أمام المستندات المستلمة .

مدة براءة الاختراع (عشرون سنة) ، و مدة شهادة المنفعة (سبع سنوات) ، ويجب سداد الرسم السنوي في بداية كل سنة اعتبار من السنة التالية لتاريخ تقديم الطلب و بانتظام .